STUDENT SOLUTIONS MANUAL AND STUDY GUIDE FOR

TECHNICAL

COLLEGE
PHYSICS

THIRD EDITION

BY JERRY D. WILSON

JOHN L. KINARD
Lander College

SAUNDERS COLLEGE PUBLISHING
Harcourt Brace Jovanovich College Publishers
Fort Worth Philadelphia San Diego New York Orlando Austin
San Antonio Toronto Montreal London Sydney Tokyo

Wilson: Student Solutions Manual and Study Guide to Accompany
TECHNICAL COLLEGE PHYSICS, 3RD EDITION

ISBN 0-03-074586-1

234 021 987654321

About the Study Guide

The Study Guide is an aid to help the student be successful in physics. Proper use of the study guide should aid in the student understanding of physics. The book includes sample problems by sections. These problems will give you more practice and some more examples as you solve problems. The book also includes approximately twenty percent of the problems in the text solved. This will allow the student to check his (her) work as problems are worked. The book also includes questions on chapter materials and their answers. After you have studied and solved problems carefully in the text and study guide, take the sample quiz at the end of each chapter. Then check your answers with the solutions in the back of the book.

Good luck as you study physics !

Table of Contents

Chapter 1   Measurement and Systems of Units

Sample Problems

What We Measure and Units

Example 1 A football player weighs 250 lb on Earth. What would his weight be on the moon ?

Solution Since the acceleration due to gravity on the moon is (1/6) the acceleration due to gravity on the earth, the weight of the player would be (1/6) of his weigh on Earth.

250 lb / 6 = 42 lb

Example 2 Each side of a solid cube has a length of 10.0 cm. The mass of the cube is measured to be 5.00×10^2 g.
A. Find the volume of the cube.
B. Find the density of the cube.

Solution A. The volume of a cube can be found by the following :
$$V = s^3$$
$$V = (10.0 \text{ cm})^3 = 1000 \text{ cm}^3$$
B. density = m / V
$$\text{density} = (5.00 \times 10^2 \text{ g}) / 1000 \text{ cm}^3 = 0.500 \text{ g /cm}^3$$

Example 3 Water has a density of 1.00×10^3 kg/m^3. Find the mass of water needed to fill a pool which is
A. circular (radius = 4.0 m and depth 1.5 m).
B. rectangular (8.0 m by 5.0 m and depth 1.5 m).

Solution   A. First find the volume of the pool.

$$V = \pi r^2 l$$

$$V = \pi (4.0 \text{ m})^2 (1.5 \text{ m}) = 75 \text{ m}^3$$

density $= m / V$

$$1000 \text{ kg} / \text{m}^3 = m / 75$$

$$m = 75,000 \text{ kg}$$

B. Again find the volume of the pool.

$$V = l \times w \times h$$

$$V = (8.0 \text{ m}) (5.0 \text{ m}) (1.5 \text{ m}) = 60 \text{ m}^3$$

density $= m / V$

$$1000 \text{ kg} / \text{m}^3 = m / 60 \text{ m}^3$$

$$m = 60,000 \text{ kg}$$

Conversions Factors

Example 1   Convert the following units of length

A. 4.0 mi = _____ km

B. 7.0 ft = _____ m

C. 2 ft 3 in = _____ cm

Solution

A.   (4.0 mi) (1 km / 0.6214 mi) = 6.4 km

B.   (7.0 ft) (1 m / 3.281 ft) = 2.1 m

C.   (2 ft)(12 in / 1 ft) = 24 in ; 24 in + 3 in = 27 in
       (27 in) (2.54 cm / 1 in) = 69 cm

Example 2   Convert the following.

A. 2 days = _____ s

B. 160 lb = _____ kg

C. 4.0 mi^2 = _____ ft^2

D. 500 cm^3 = _____ m^3

Solution

A.   (2 days) (24 h / 1 day) (3600 s / 1 h) = 172,800 s

B.   (160 lb) (1 kg / 2.2 lb) = 73 kg

C.   (4.0 mi^2) (5280 ft / 1 mi)^2 = 1.1 x 10^8 ft^2

D.   (500 cm^3) (1 x 10^{-6} m^3 / 1 cm^3) = 5.0 x 10^{-4} m^3

Example 3   A home has 1950 ft^2 of heated floor space. Find the area in m^2.

Solution   (1950 ft^2) (1 m^2 / 10.76 ft^2) = 181 m^2

Example 4   Convert 55 mi/h to km/h and into m/s

Solution   (55 mi / h) (1 km / 0.6214 mi) = 88.5 km /h

               (88.5 km/h) (1000 m/km) (1 h / 3600 s) = 24.6 m/s

Solutions to Selected Problems from the Text

4. (a) A = 4 π r^2

       A = 4 π (0.10)^2 (diameter is one half the radius of the sphere)

       A = 0.13 m^2

  (b) V = (4/3) πr^3

       V = (4/3) π (0.10)^3

       V = 0.0042 m^3

8. given : radius of the sphere = 8.0 cm ; mass = 1000 g
First find the volume of the sphere.

$V = (4/3)\pi r^3$

$V = (4/3)\pi (8.0)^3 = 2100 \text{ cm}^3$

density = m / V

density = $(1000 \text{ g}) / 2100 \text{ cm}^3 = 0.47 \text{ g/cm}^3$

14. given : density of mercury 13.6 g/cm^3 ; mass = 68 g ; find V

   density = m/V
   13.6 = 68 / V
   V = 68 / 13.6
   V = 5.0 cm^3

19. (a) (20 mi) (1.609 km / 1 mile) = 32 mi
    (b) (10.0 ft) (30.48 cm / 1 ft) = 305 cm
    (c) (30 in) (1 m / 39.37 in) = 0.76 m
    (d) (500 yd) (3 ft / 1 yd) (3.048 x 10^{-4} km / ft) = 0.457 km

24. 1 ton = 2000 lb
    1 metric ton = 1000 kg
    (1000 kg) (2.2 lb / kg) = 2200 lb
    The metric ton is larger.

29. (12 people) (1/4 lb / serving) = 3.0 lb
    (3.0 lb)(1 kg / 2.2 lb) = 1.4 kg

32. (100 ft^2) (1 yd / 3 ft)^2 (notice you must square the conversion factor)

    11.1 yd^2

36. set up a proportion

   $\dfrac{(0.25 \text{ in})}{(2.5 \text{ in})} = \dfrac{(1 \text{ ft})}{x \text{ ft}}$ $\dfrac{(0.25 \text{ in})}{(3.0 \text{ in})} = \dfrac{(1 \text{ ft})}{x \text{ ft}}$

   (0.25) x = 2.5   (0.25) x = 3.0
   x = 10 ft   x = 12 ft

   Find the area for the rectangle : A = l x w

   A = (10 ft)(12 ft) = 120 ft^2

4

Review Questions

1. Distinguish between mass and weight.

   Mass is the quantity of an object contains and is a fundamental property. Weight is the force of gravity on an object by some celestial body, usually Earth, and varies for different bodies like the Earth and moon.

2. What is density ?

   The compactness of matter is expressed in terms of density, and the mass (or weight) per unit volume ($\rho = m/V$, mass density, or $D = w/V$, weight density).

3. What is a standard unit ?

   A standard unit is a fixed and reproducible value for the purpose of taking accurate measurement. Examples include the foot and meter.

4. What are the numerical bases for the metric and British systems of units ?

   The metric system is a decimal or base-10 system. The British system is a duodecimal or base-12 system.

5. How many times longer is a kilometer than a centimeter ?

   Since $1\ m = 10^2$ cm and $1\ km = 10^3$ m, a kilometer is 10^5 times as long as a centimeter, or $1\ km = 10^5$ cm.

6. Explain why water has a density of 1.0 in units of g/cm^3.

   The kilogram (1000 g) was originally defined as the quantity of mass continued in a volume of water 1 dekameter (10 cm) on a side.

5

$$\rho = m/V = 1000 \text{ g} / 1000 \text{ cm}^3 = 1.0 \text{ g/cm}^3$$

7. What are the standard units of mass in the SI system and the British system ?

The kilogram and the slug, respectively. The slug is defined in terms of force or weight, and 1 slug has a weight of $w = mg = (1)(32) = 32$ lb.

8. How do units help in the use of conversion factors ?

The units "cancel" in conversion operations so as to let you know the use is correct - whether to multiply or divide. An equation must by not only equal in magnitude on both sides, but also equal in units.

9. What are two important initial steps in solving a problem ?

Understanding what the problem in knowing <u>what is given</u> and <u>what is wanted</u>. Then, one decides how to work the problem.

10. How many significant figures or digits (numbers) should be reported when solving a problem ?

Technically, a result should not have no more figures than the quantity used in the calculation with the least number of significant figures. Sometimes more significant figures are assumed and reported as a necessity for clarification.

Sample Quiz

(Remove the quiz from the book and test your knowledge of the chapter material as though you were taking an in-class quiz. Check your answers with the key at the back of the Study Guide.)

Completion

1. _____ describes an object's size and specifies its position in space.

2. The metric prefix _____ means one-millionth.

3. The standard unit of mass in the SI system is the _____.

Multiple Choice

____ 4. Which of the following has the largest volume ?
   A. cubic dekameter
   B. 500 mL
   C. L
   D. quart

____ 5. In a metric country, instead of buying 10 lb of sugar you would most nearly but what quantity of sugar ?
   A. 1 kg   B. 1 slug   C. 5 kg   D. 10 kg

Problems

6. A visiting foreign student lists his height and weight as 160 cm and 60 kg, respectively. What are his height and weight in the British system ?

7. A container holds 5.00 kg of water. Find the volume of the water in
   A. liters and
   B. cubic feet

7

Chapter 2   Technical Mathematics

Sample Problems

Scientific Notation

Example 1   Write the following numbers in scientific notation.

    A. 9,000,000,000   _____
    B. 3450   _____
    C. 2860.3   _____
    D. 0.0062   _____
    E. 0.000000000067   _____

Remember the rule - one number before the decimal and the rest behind the decimal.

Solution

    A. 9×10^9
    B. 3.45×10^3
    C. 2.8603×10^3
    D. 6.2×10^{-3}
    E. 6.7×10^{-11}

Example 2   Complete the following.

    A. $(3.0 \times 10^8)(2.0 \times 10^{-6})$ _____
    B. $(4.0 \times 10^3)(6.0 \times 10^{-2})$ _____
    C. $(1.0 \times 10^2) / (5.0 \times 10^6)$ _____

Solution   A. $3 \times 2 = 6$; add the exponents 6.0×10^2

               B. $4 \times 6 = 24$; add the exponents $24 \times 10^1 = 2.4 \times 10^2$

               C. re-write 10×10^1 then divide 10 by 5 = 2 and subtract
                    the exponents 10^{-4} 2×10^{-4}

Example 3 Complete the following.

A. $(4.0 \times 10^{-3}) / (2.0 \times 10^{-6})$ _____

B. $(1.8 \times 10^{-3}) / (6.0 \times 10^{-2})$ _____

C. $(1.0 \times 10^{2}) / (5.0 \times 10^{6})$ _____

Solution A. $4/2 = 2$; the exponents $-3 - (-6) = 3$ 2.0×10^{3}
B. $18 \times 10^{-4} / 6.0 \times 10^{-2}$ 3.0×10^{-2}
C. $10 \times 10^{1} / 5.0 \times 10^{6}$ 2.0×10^{-5}

Angle Measure and Trigonometry

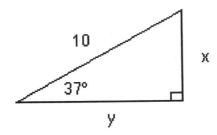

Example 1

Find x and y in the diagram above.

Solution $\sin 37 = x / 10$
$0.6 = x / 10$
$x = 6$

$\cos 37 = y / 10$
$0.8 = y / 10$
$y = 8$

9

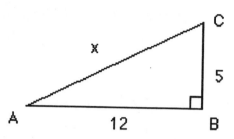

Example 2   In the diagram,
A. find x
B. find the measure of angle A

Solution   By the Pythagorean theorem $x^2 = 12^2 + 5^2$
$$x^2 = 144 + 25$$
$$x^2 = 169$$
$$x = 13$$

Tan A = 5 / 12
Tan A = 0.42
A = 23°

Example 3   A wheel makes ten complete revolutions.
A. Find the number of degrees in ten revolutions.
B. Find the number of radians in ten revolutions.

Solution   A. (10 revolutions) (360° / revolution) = 3600°
B. (10 revolutions)(2 π radians / revolution)=20 π =63 rad

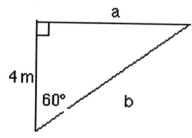

Example 4

Find a and b.

10

Solution $\cos 60 = 4 / b$ $b^2 = a^2 + 4^2$

        $0.5 \quad = 4 / b$ $8^2 = a^2 + 4^2$

        $b \quad = 4 / 0.5$ $64 = a^2 + 16$

        $b \quad = 8.0\,m$ $a \quad = 6.9\,m$

Vectors

Example 1   Find $\mathbf{A} + \mathbf{B}$ if $A = 6\,m\,+\mathbf{x}$ and $B = 8\,m\,-\mathbf{y}$.

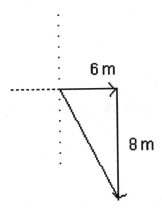

6 m

8 m

Solution

        Find $\mathbf{r}$ by using the Pythagorean theorem.

        $r^2 = 6^2 + 8^2$ $\mathrm{Tan}\,\theta = 8 / 6$

        $r = 10\,m$ $\theta = 53°$ below the $+ x$ axis

Example 2   Find the the x and y components if $\mathbf{r} = 10\,m$ and $\varnothing = 143°$

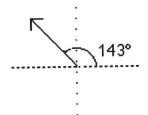

143°

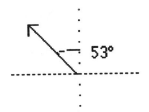

53°

or

$x = r\ \sin 53$ $y = r\ \cos 53$

$x = 10(\sin 53)$ $y = 10(\cos 53)$

$x = -8.0\,m$ $y = + 6.0\,m$

11

Example 3   Find $A + B$ if $A = 6$ m $+x$ and $B = 6$ m $+50°$

Solution

First break each vector into components.

$A_x = 6$ m $A_y = 0$

$B_x = 6$ m $(\cos 50) = 3.9$ m $B_y = 6$ m $(\sin 50) = 4.6$ m

$R_x = 9.9$ m $R_y = 4.6$ m

$R^2 = (9.9)^2 + (4.6)^2$ (since the components are
perpendicular)

$R = 10.9$ m

$\text{Tan}\, \theta = 4.6 / 9.9$

$\theta = 25°$

Example 4   Find $A + B + C$ if $A = 4.0$ m east , $B = 6.0$ m $30°$ S of W ,
and $C = 8.0$ m west.

Solution

Break each of the vectors into components.

$A_x = 4.0$ m E $A_y = 0$

$B_x = 6.0$ m $(\cos 30) = 5.2$ m W $B_y = 6.0(\sin 30) = 3.0$ m S

$C_x = 8.0$ m west $C_y = 0$

$R_x = 9.2$ m west $R_y = 3.0$ m south

$R^2 = 9.2^2 + 3.0^2$

$R = 9.7$ m

$\text{Tan}\, \theta = 3.0 / 9.7$

$\theta = 17°$ south of west

Solutions to Selected Problems from Text

4. (a) $(2.0 \times 10^2)(6.0 \times 10^3) = 12 \times 10^3$ (multiply 2 and 6 - then add the exponents)

$$= 1.2 \times 10^4$$

(b) $(3.5 \times 10^4)(2.0 \times 10^{-3}) = 7.0 \times 10^1$ (multiply 2 and 3.5 - then add the exponents)

(c) $(5.0 \times 10^{-4})(3.2 \times 10^{-2}) = 16 \times 10^{-6}$ (multiply 5 and 3.2 - then add the exponents)

(d) $(1.0 \times 10^{-5})(8.6 \times 10^{12}) / 2.0 \times 10^4 = 4.3 \times 10^3$

(divide 8.6 / 2 = 4.3)

$-5 + 12 - 4 = 3$

9. (a) $(4.32 \times 10^4) + (2.5 \times 10^3) = (43.2 \times 10^3 + 2.5 \times 10^3) = 45.2 \times 10^3$

$$= 4.52 \times 10^4$$

(b) $(9.2 \times 10^7) + (6.0 \times 10^8) = 9.2 \times 10^7 + 60 \times 10^7 = 69.2 \times 10^7$

$$= 6.92 \times 10^8$$

(c) $(1.2 \times 10^{-5}) + (3.5 \times 10^{-4}) = 0.12 \times 10^{-4} + 3.5 \times 10^{-4} = 3.62 \times 10^{-4}$

(d) $(6.6 \times 10^8) - (1.1 \times 10^8) = 5.5 \times 10^8$

(e) $(4.0 \times 10^5) - (2.0 \times 10^4) = (40 \times 10^4 - 2.0 \times 10^4) = 38 \times 10^4$

$$= 3.8 \times 10^5$$

(f) $(9.5 \times 10^{-2}) - (4.0 \times 10^{-3}) = (9.5 \times 10^{-2}) - (0.4 \times 10^{-2})$

$$= 9.1 \times 10^{-2}$$

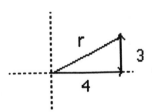

14.

   (a) $r^2 = 3^3 + 4^2$

        $r^2 = 25$

        $r = 5.0 \, m$

   (b) $\sin \theta = 3/5 = 0.60$

        $\cos \theta = 4/5 = 0.80$

        $\tan \theta = 3/4 = 0.75$

(c) $\sin \varnothing = 4 / 5 = 0.80$
       $\cos \varnothing = 3 / 5 = 0.60$
       $\tan \varnothing = 4/3 = 1.33$

(d) $\theta = \text{arc } (\sin 0.60) = 37° \ (1 \text{ rad} / 57.3°) = 0.65 \text{ rad}$
       $\varnothing = \text{arc } (\sin 0.80) = 53° \ (1 \text{ rad} / 57.3°) = 0.92 \text{ rad}$

19. $\sin 60 = y / 10 \text{ m}$
    $y = (\sin 60)(10 \text{ m}) = 8.7 \text{ m}$
    $\cos 60 = x / 10 \text{ m}$
    $x = (\cos 60)(10 \text{ m}) = 5.0 \text{ m}$

26. $y = 3.6 \text{ in} ; r = 1.0 \text{ ft}$
    $\text{slope} = \text{rise} / \text{run} = 3.6 / 12 = 0.30 = 30\%$
    $\text{Tan } \varnothing = 3.6 \text{ in} / 12 \text{ in} = 0.30$
    $\varnothing = 17°$

29.   since $\theta = 45°$; $\varnothing = 45°$; the triangle is isosceles and side A = side
    B = 12 m

    $\cos \theta = 2 / C$
    $C = 12 / \cos 45 = 17 \text{ m}$

34. $\cos 37 = A / 2.0$
    $0.8 = A / 2.0$
    $A = 1.6 \text{ m}$
    $\sin 37 = B / 2.0$
    $0.6 = B / 2.0$
    $B = 1.2 \text{ m}$

39.   **A** = 10.0 m 30° to the + x axis ; **B** = 12.0 m at an angle of 70° + x
    A : $x = 10.0 (\cos 30) = + 8.7$ $y = 10.0 (\sin 30) = +5.0$
    B : $x = 12.0 (\cos 70) = + 4.1$ $y = 12.0 (\sin 70) = 11.3$
           $x = 12.8$ $y = 16.3$

        $r^2 = x^2 + y^2$
        $r^2 = 12.8^2 + 16.3^2$
        $r = 20.7 \text{ m}$
        $\text{Tan } \theta = 16.3 / 12.8$
        $\theta = 52°$ with the positive x-axis

44. F_1 : $F_{1x} = 6.0 \,(\cos30) = 5.2$ N $F_{1y} = 6.0 \,(\sin30) = 3.0$ N
     F_2 : $F_{2x} =$ 0 $F_{2y} =$ 4.0 N
     F_3 : $F_{3x} = 8.0 \,(\cos45) = -5.7$ N $F_{3y} = 8.0 \,(\sin45) = 5.7$ N
     F_4 : $F_{4x} = 6.0 \,(\cos60) = -3.0$ N $F_{4y} = 6.0 \,(\sin60) = -5.2$ N

Now find the total x and y

$$F_x = 5.2 - 5.7 - 3.0 = -3.5 \text{ N}$$
$$F_y = 3.0 + 4.0 + 5.7 - 5.2 = 7.5 \text{ N}$$
$$F_r^2 = 3.5^2 + 7.5^2$$
$$F_r = 8.3 \text{ N}$$

$\text{Tan}\,\theta = 7.5 / -3.5$

$\theta = 65°$ with respect to the -x - axis

Review Questions

1. If the decimal point of a number in powers of ten notation is shifted, how does his affect the exponent or power of ten ?

   The exponent is decreases by one for every place the decimal is shifted to the right and increased for every place shifted to the left.

2. What is the standard form of expressing a number in powers of ten notation ?

   Having one number or digit to the left of the decimal point.

3. What is necessary when taking the square root of a number in scientific notation ?

   The exponent of the power of ten must be even or divisible by two.

4. How many times greater is the circumference of a circle than its radius ?

15

Since $\theta = s/r = 2\pi$ (or there are 2π rad in one circle or revolution), . $s = 2\pi r = 6.28$ times greater. Also $c = \pi d = 2\pi r$.

5. What is the ratio of the side opposite an angle to the hypotenuse of a right triangle ?

opposite / hypotenuse = $\sin \theta$

6. What is meant by the rise over the run ?

The ratio of the vertical distance to the horizontal distance gives the slope of an incline or the tangent of the angle of the incline.

7. What is the difference between a scalar quantity and a vector quantity ?

A scalar is a numerical quantity with magnitude only.
A vector has both magnitude and direction.

8. In vector addition, what is the sum or two or more vectors called ?

The vector sum or resultant.

9. For a general vector **F**, what are its x and y components ?

The projections of **F** on the x and y axes, generally,
$\mathbf{F}_x = \mathbf{F}(\cos\varnothing)$ and $\mathbf{F}_y = \mathbf{F}(\sin\varnothing)$.

10. What is the direction of a vector in terms of its components ?

$\varnothing = \tan^{-1}(\mathbf{F}_y / \mathbf{F}_x)$, relative to the x-axis.

Sample Quiz

(Remove the quiz from the book and test your knowledge of the chapter material as though you were taking an in-class quiz. Check your answers with the key at the back of the Study Guide.)

Completion

1. The numbers 8400 and 0.00016 written in scientific notation are
_____ and _____.

2. An angle of 90° is equivalent to _____ radians.

3. Given the angle relative to the x-axis and the x-side of a right triangle, the length of the y-side of the triangle is given by_____.

Multiple Choice

____4. The $\cos(90° - \emptyset)$ for a right triangle is equal to
A. $\cos\emptyset$ B. $\sin\emptyset$ C. $\sin(90° - \emptyset)$ D. $\cos 90$

____5. For a right triangle, $\tan\emptyset$ is equal to
A. side opposite over the hypotenuse
B. side adjacent over side opposite
C. $(\cos\emptyset - \sin\emptyset)$
D. $\sin\emptyset / \cos\emptyset$

Problems

6. A particle in a circle of radius 0.40 m through an angle of 210°. What is the arc length traveled by the particle ?

7. Given three vectors : $F_1 = 10$ N at $\emptyset = 30°$
$F_2 = 8.0$ N at $\emptyset = 180°$
$F_3 = 20$ N at $\emptyset = 300°$
What is the vector sum of theses three vectors ?

Chapter 3   Statics and Equilibrium

Sample Problems

Statics and Free Body Diagrams

Example 1   Given the two vectors below. Find if the two vectors are in equilibrium and if they are not, find the force which would place the system in equilibrium.

$$F_1 = (8\,x - 3\,y)\ N$$
$$F_2 = (-2\,x + 4y)\ N$$

Solution

Work with the x and y components of the vectors separately from each other.

The sum of the x- components is **6x** and the sum of the y-components is **1y**. Therefore the system is not in equilibrium. To get the system in equilibrium the sum of all forces in x and y should be zero.

$$F_{eq} = (-6x - y)\ N$$

Example 2   In the diagram find T_1 , T_2 , and T_3.

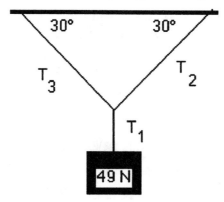

Solution

T_1 must support the weight of the mass, therefore $T_1 = 49$ N. Use the point where the three mass are connected as the reference point.

F_x $\quad\quad\quad$ $0 = T_2 \cos 30 - T_3 \cos 30$

from this

$T_2 = T_3$

F_y $\quad\quad\quad$ $0 = T_2 \sin 30 + T_3 \sin 30 - T_1$

$T_1 = 2\, T_2 \sin 30$ $\quad\quad$ (since $T_2 = T_3$)

$T_1 = T_2 = 49$ N

$T_3 = 49$ N

Example 3 $\quad$ Find T_1, T_2, and T_3 in the diagram.

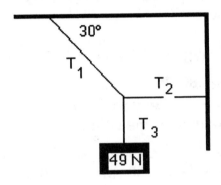

Solution

$T_3 = 49$ N since it is a vertical force supporting the weight.

F_x $\quad\quad\quad$ $0 = T_1 \cos 30 - T_2$ $\quad\quad$ (two unknowns)

F_y $\quad\quad\quad$ $0 = T_1 \sin 30 - T_3$

$0 = T_1 (0.5) - 49$ N

$T_1 = 98$ N

$0 = (98 \text{ N}) \cos 30 - T_2$ $\quad\quad$ substitute into F_x

$T_2 = 85$ N

Rigid Body Statics

Example 1   A mass of 200 g is placed on the 30 cm position of a meter stick and a 400 g mass is placed on the 80 cm position. Neglecting the mass of the stick, find the force and the location of the force to balance the meter stick and its weights.

<u>Solution</u>

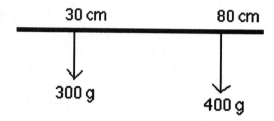

30 cm   80 cm

300 g

400 g

The total of the upward forces must be equal to the total downward forces.

$$F_{up} = (0.300 \text{ kg})(9.80 \text{ m/s}^2) + (0.400 \text{ kg})(9.80 \text{ m/s}^2)$$
$$F_{up} = 6.86 \text{ N}$$

To find the location where the force should be applied, choose the end of the meter stick.

$$\tau_{clockwise} = \tau_{counterclockwise}$$
$$F_{up} \, x = (2.94) \, (30 \text{ cm}) + (3.92)(80 \text{ cm})$$
$$6.86 \, (x) = 402$$
$$x = 58.6 \text{ cm from the 0 cm mark}$$

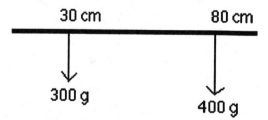

30 cm   80 cm

300 g

400 g

Example 2

In the diagram above, suppose each end of the meter stick were supported by spring scales. Find the reading of each spring scale.

F_1 - left end and F_2 - right end.

$F_1 + F_2 = (0.300 \text{ kg}) (9.80 \text{ m/s}^2) +$

$(0.400 \text{ kg})(9.80 \text{ m/s}^2) = 6.86 \text{ N}$

Set the pivot or reference point on the left end of the stick.

$\tau_{clockwise} = \tau_{counterclockwise}$
$F_2 (100 \text{ cm}) = (2.94)(30 \text{ cm}) + (3.92)(80 \text{ cm})$
$F_2 (100) = 401.8$
$F_2 = 4.02 \text{ N}$

$F_1 = 6.86 \text{ N} - 4.02 \text{ N} = 2.84 \text{ N}$

Solutions to Selected Problems from the Text

4. $F_1 = 4x - 2y + 3z$
   $F_2 = 3x + 5y - 7z$
   $F_3 = -7x - 3y + 4z$
   $F_r = 0x + 0y + 0z = 0$. The system is in equilibrium.

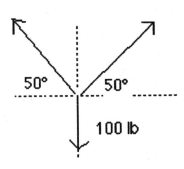

100 lb

7.

The tension in the lower cable is 100 lb.
Since the body is in equilibrium, the net force is 0.
F_x : $0 = T_1 \cos 50 - T_2 \cos 50$ T_1 left cable ; T_2 right cable
     $T_1 = T_2$

F_y: $\quad 0 = T_1 \sin 50 + T_2 \sin 50 - 100 \text{ lb}$

$\qquad 2\, T_1 \sin 50 = 100 \text{ lb}$

$\qquad T_1 = 50 / \sin 50 = 65 \text{ lb}$

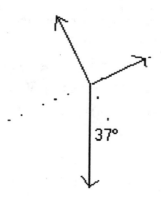

12.

F_x (parallel to the plane) $\qquad 0 = mg \sin 37 - f_s$

$\qquad\qquad\qquad\qquad\qquad\qquad 0 = mg \sin 37 - 50.0 \text{ N}$

$\qquad\qquad\qquad\qquad\qquad\qquad mg = 50 \text{ N} / \sin 37 = 83 \text{ N}$

F_y (perpendicular to the plane) $\qquad 0 = N - mg \cos 37$

$\qquad\qquad\qquad\qquad\qquad\qquad\qquad N = 83 \text{ N} (\cos 37) = 67 \text{ N}$

17. The downward forces total 75 N . The upward force = 50 N. The normal force should be the difference between the upward and downward forces. The normal force equals 25 N.

23. $\tau = Fr \sin \varnothing$

$\quad \tau = (10 \text{ lb})(4.0 \text{ ft}) (\sin 30) = 20 \text{ ft-lb}$

27. $\tau_{\text{clockwise}} = \tau_{\text{counterclockwise}}$

$\quad (20 \text{ N })(40 \text{ cm}) = (15 \text{ N})(x) \qquad$ x is measured from the right of 50 cm

$\quad (20)(40) / (15) = x$

$\quad x = 53$ cm from the point of 50 cm ; 103 cm from 0 - therefore it is off the stick. If the pivot were changed, the object could be placed in equilibrium.

31. The total force up must equal the total downward force.

$F_A + F_B = 50,000$ lb

Now work with torques. Choose the left end as the pivot.

τ clockwise $= \tau$ counterclockwise

(a) $(40,000)(50) + (10,000)(50) = F_B (100)$

$F_B = 25,000$ lb   ; $F_A = 25,000$ lb

(b) $(10,000)(50) + (40,000)(90) = F_B(100)$

$F_B = 41,000$ lb   ; $F_A = 9,000$ lb

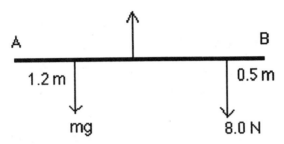

37.

clockwise torque equals the counterclockwise torque

$(8.0\ N)(1.5\ m) = (mg)(0.8\ m)$
$mg = 15\ N$

Review Questions

1. What is a particle ?

Theoretically, an amount of matter assumed to occupy a single point in space. A particle has no dimensions so rotational motion is not a consideration.

2. What is necessary for particle static equilibrium ?

That the resultant force acting on the particle at rest is zero.

3. Distinguish between a space diagram and a free body diagram.

A space diagram is a sketch of the actual situation. A free-body diagram illustrates an object as a particle and shows the concurrent forces acting on it.

4. Distinguish between concurrent and coplanar forces.

Concurrent forces act through a common point. Coplanar forces are in the same plane. Coplanar forces can be concurrent, but all concurrent forces are not coplanar.

5. If a block sits on an inclined plane, what is the weight component of the block down or parallel to the plane ?

$F = w \sin \emptyset = mg \sin\emptyset$

6. Explain how two equal and opposite forces can have a zero resultant but still produce no motion.

If the forces are not concurrent, there will be a torque which can produce rotational motion.

7. What is the condition for static rotational equilibrium ?

That the sum of the torques, or the resultant torque, be zero.

8. Why does it make no difference where one chooses an axis of rotation in a static system to sum the torques ?

If a system is in static equilibrium, the sum of the torques will be zero (no rotational motion) at any point in the system.

9. Distinguish between the center of gravity and center of mass for a rigid body.

The center of gravity is the point at which the entire weight (or mass) of an object may be considered to be concentrated. It is

the point about which the sum of the torques is zero. In a uniform gravitational field, the center of mass coincides with the center of gravity.

10. What causes an object in unstable equilibrium to be unstable ?

When slightly displaced about the object's axis or point of support, the center of gravity is such that it supplies a torque that causes the object to rotate or topple.

Sample Quiz

(Remove the quiz from the book and test your knowledge of the chapter material as though you were taking an in-class quiz. Check your answers with the key at the back of the Study Guide.)

Completion

1. _____ forces act through a common point.

2. Torque is equal to the product of _____ and _____.

3. A rigid body is not in rotational equilibrium when there is a net _____ acting on the body.

Multiple Choice.

___4. In a free-body diagram,
A. the forces always act through the center of gravity
B. an object is represented as a rigid body
C. the forces act through a common point
D. the forces must be coplanar

___5. When an object in stable equilibrium is slightly displaced, its center of gravity is
A. shifted to another location in the body
B. raised
C. outside the base of support
D. none of the above

Problems

6. Two forces $F_1 = (3\,\mathbf{x} - 4\,\mathbf{y})$ N and $F_2 = (-5\,\mathbf{x} + 3\,\mathbf{y})$ N , act on a
   particle. What third force will put the particle in static equilibrium ?

7. A meter stick is pivoted at the 30 cm position. If a weight of 5.0 N is
   suspended from the 10 cm position, where would a 3.0 N weight
   have to be suspended so that the stick is in static equilibrium ?

Chapter 4 Motion : Description and Analysis

Sample Problems

Motion, Speed, and Velocity

Example 1 A car travels 55 mi/h for a total time of 4.5 h. How far did the car travel ?

Solution

given: $v = 55$ mi/h $\qquad$ $t = 4.5$ h $\qquad$ find d

$$v = x / t$$
$$55 = x / 4.5$$
$$(55)(4.5) = x$$
$$2.5 \times 10^2 \text{ mi} = x$$

Example 2 An automobile travels 30 m/s East for 45 minutes then 20 m/s West for 30 minutes.
A. Find the car's average speed for trip.
B. Find the car's average velocity.

Solution

Break the problem down in two parts - the segments of the trip.

$v_1 = x_1 / t$ $\qquad\qquad\qquad$ $v_2 = x_2 / t$

$30 = x_1 / 2.7 \times 10^3$ $\qquad\qquad$ $20 = x_2 / 1.8 \times 10^3$

$x_1 = 8.1 \times 10^4$ m East $\qquad\qquad$ $x_2 = 3.6 \times 10^4$ m West

(a) the total distance traveled is 11.7×10^4 m in a time of 4.5×10^3 s

$$v = x / t$$
$$v = 11.7 \times 10^4 / 4.5 \times 10^3 = 26 \text{ m/s}$$

28

(b) the total displacement is 4.5×10^4 m in a time of 4.5×10^3 s

$$v = x / t$$
$$v = 4.5 \times 10^4 / 4.5 \times 10^3$$
$$v = 1.0 \times 10^1 \text{ m/s East}$$

Acceleration

Example 1   A car accelerates from rest to 30 m/s in 7.5 seconds. Find the car's acceleration.

Solution

$v_o = 0$ $v_f = 30$ m/s $t = 7.5$ s $a = ?$

$$a = (v_f - v_o) / t$$
$$a = (30 - 0) / (7.5)$$
$$a = 4.0 \text{ m/s}^2$$

Example 2   An airplane accelerates from rest to 60.0 m/s as it travels 1.0×10^3 m.
A. Find the acceleration of the airplane.
B. Find the speed of the plane after 10.0 s
C. How long will it take the plane to take off ?

Solution

$v_o = 0$ $v_f = 60.0$ m/s $x = 1.0 \times 10^3$ m $a = ?$

(a) $v_f^2 = v_o^2 + 2 ax$

$$(60.0)^2 = (0)^2 + 2 \, (a)(1.0 \times 10^3)$$
$$(3.6 \times 10^3) / 1.0 \times 10^3 = 3.6 \text{ m/s}^2$$

(b) $v_f = v_o + at$

$$v_f = 0 + (3.6) \, (10)$$
$$v_f = 36 \text{ m/s}$$

(c)   from part b

$$60 = 0 + (3.6)(t)$$
$$60 / 3.6 = t$$
$$t = 17 \text{ s}$$

29

Example 3 A motorist traveling at 20.0 m/s applies his brakes for a
distance of 20.0 m.   After this distance the motorist is
traveling at 10.0 m/s.
A. How long does the motorist apply the brakes ?
B. What is the acceleration of the car ?
C. What additional time is needed for the car to come to a
stop ?

Solution

Here is a problem where it may be easier to find part b
then a and c

B. $v_0 = 20$ m/s $x = 25$ m $v_f = 15$ m/s   find a

$$v_f^2 = v_0^2 + 2ad$$

$$(15)^2 = (40)^2 + 2(a)(25)$$
$$(225 - 400) / 50 = a$$
$$a = -3.5 \text{ m/s}^2$$

A. $v_0 = 20$ $v_f = 15$ $a = -3.5$ m/s^2 find t
    $$v_f = v_0 + at$$
    $$15 = 20 - (3.5)t$$
    $$t = 1.4 \text{ s}$$

C. $v_0 = 15$ $v_f = 0$ $a = 3.5$ m/s^2 find t
    from (A) $0 = 15 + (-3.5)(t)$
                    $t = 4.3$ s

Free Fall

Example 1 A rock is dropped from the top of a building which is 78 m
tall.
A. Find the time the rock is in the air.
B. Find the velocity of the rock as it strikes the ground.

A. $v_0 = 0$ $g = 9.8$ m/s^2 $y = -78$ m   find t

$$y = v_0 t - (1/2) gt^2$$

30

$$-78 = 0 - (1/2)(9.8)(t^2)$$
$$-78 / -4.9 = t^2$$
$$t = 4.0 \text{ s}$$

B. $v_f = v_0 - gt$ **or** $v_f^2 = v_0^2 - 2gy$

$\quad v_f = 0 - (9.8)(4.0)$ $v_f^2 = 0^2 - 2(9.8)(-78)$

$\quad v_f = -39 \text{ m/s}$ $v_f = -39 \text{ m/s}$

Example 2   A stone is thrown into the air and attains a height of 20.0 m before it is caught at the same height as it was thrown.
A. With what speed was the ball thrown ?
B. How long was the ball in the air ?

Solution

A. $g = 9.8 \text{ m/s}^2$ $v_{top} = 0$ $y = 20.0 \text{ m}$ find v_0

$\qquad v_f^2 = v_0^2 - 2gy$

$\qquad 0^2 = v_0^2 - 2(9.8)(20)$

$\qquad v_0 = 19.8 \text{ m/s}$

B.   using information in part A

$\qquad v = v_0 - gt_{up}$

$\qquad 0 = 19.8 - (9.8)(t)$

$\qquad t_{up} = 2.02 \text{ s}$

Since object is caught at same height as it was thrown
$t_{total} = 4.04 \text{ s}.$

Example 3   A mass is thrown upward at 10.0 m/s from the top of a building 50.0 m tall. A gentle wind slightly moves the mass while it is in flight and it hits the ground.
A. How high above the ground will the mass ascend ?
B. With what velocity will the mass strike the ground ?
C. How long is the mass in flight ?

Solution

A. $v_0 = 10.0$ m/s $g = 9.80$ m/s^2 $v_{top} = 0$ find y

$$v_f^2 = v_0^2 - 2gy$$
$$0^2 = 10^2 + 2(-9.8)(y)$$
$$100 / -19.6 = y$$
$$y = 5.1 \text{ m} \quad \text{therefore mass is 55.1 m from the ground}$$

B. $v_0 = 10.0$ $g = -9.80$ m/s^2 $y = -50.0$ m   find v_f

$$v_f^2 = v_0^2 - 2gy$$
$$v_f^2 = 10^2 - 2(9.8)(-50)$$
$$v_f = -32.9 \text{ m/s}$$

C. $v_f = v_0 - gt$
$$-32.9 = 10 - (9.8) t$$
$$-42.9 / -9.8 = t$$
$$t = 4.38 \text{ s}$$

Solutions to Selected Problems from Text

4. given : $t = 20$ s ; $x = 100$ yd $= 300$ ft
$$v = x / t$$
$$v = 300 / 20 = 15 \text{ ft/s toward the other goal line}$$

7. (a) given : diameter of circle = 1.0 mi ; $t = 94$ s $= 2.6 \times 10^{-2}$ h
$$x = \pi d = (3.14)(1.0 \text{ mi}) = 3.14 \text{ mi}$$
$$v = x / t$$
$$v = 3.14 \text{ mi} / 2.6 \times 10^{-2} \text{ h}$$
$$v = 1.2 \times 10^2 \text{ mi/h}$$

   (b) 3.14 mi $= 1.7 \times 10^4$ ft
$$v = (1.7 \times 10^4) / 94 = 1.8 \times 10^2 \text{ ft/s}$$

10. (a) To find the average speed, find the slope of the line.
   slope = rise / run = 2.5 m / 5.0 s = 0.5 m/s

(b) Since the slope of the line is uniform for the entire graph the average velocity is the same as the instantaneous velocity. If there were a curve, the average velocity and the instantaneous velocity would not be equal.

16. (a) Since speed is a scalar the direction does not enter into the picture.

$v = x / t$

$v = 900 / 41\ s = 22\ yd / s$ (1 mi / 1760 yd) (3600 s / 1 h)=45 mi/ h

(b) First break the displacement into components :

$x_1 = (500)(\sin 37) = 300$ yd east

$y_1 = (500)(\cos 37) = 400$ yd north

$y_2 = 400$ yd south

total x = 300 yd east and total y = 0

$v_x = 300\ yd / 41\ s = (7.3\ yd / s)\ (3\ ft / 1\ yd) = 22$ ft/s

(22 ft /s)(1 mi / 5280 ft) (3600 s / 1 h) = 15 mi /h **east**

20. given : $v_0 = 12$ m/s ; $a = 8.0$ m/s^2 ; t = 6.0 s ; $v_f = ?$

$v_f = v_0 + at$

$v_f = 12 + (8.0)(6.0)$

$v_f = 12 + 48$

$v_f = 60$ m/s in the original direction of the motion

24. (A) In Figure 4.9 (b), the acceleration can be found by finding the slope of the line. Take the two points (0 s , 4.0 m/s) and (3.0 s , 0 m/s).

slope = a = (- 4.0 m/s / 3.0 s) = -1.3 m/s^2 ; the negative sign indicates the acceleration is in the opposite direction of the initial motion.

(B) The initial speed of the mass occurs when t = 0. The speed at this point is 4.0 m/s . The final speed occurs at t = 3.0 s which can be read from the graph at 0 m/s.

29. given : $v_0 = 75$ km /h ; $v_f = 15$; t = 10 s ; x = ?

$\overline{v} = (v_f + v_0) / 2 = (75 + 15) / 2 = 45$ km/h

$$x = \bar{v}\,t = (45)\,(1/6) = 7.5 \text{ km /h}$$

34. given $v_0 = 0$; $y = -60$ ft ; $g = 32$ ft/s^2 ; find v_f

$$v_f^2 = v_0^2 - 2gy$$
$$v_f^2 = 0^2 - 2\,(32)(-60)$$
$$v_f = 62 \text{ ft / s}$$

Review Questions

1. Discuss the terms used to describe motion as "time rates of change."

   Speed or velocity is the time rate of change of position.
   Acceleration is the time rate of change of velocity.

2. Which has the greater magnitude between two points, distance or displacement ?

   Distance is always greater than or equal tot he magnitude of the displacement.

3. Describe the x versus t graphs for uniform velocity and uniform acceleration.

   The graph for uniform velocity is a straight line with a positive or negative slope. If the slope were zero, no motion would be present. The graph for constant acceleration would be represented by a parabola.

4. What does negative acceleration imply ?

   In general, it means that the final velocity for a time period is less than the initial velocity, or that an object is decelerating. Another example is when the acceleration is in the opposite direction of the initial direction of motion.

5. Describe a graph of v versus t for the following situations below.

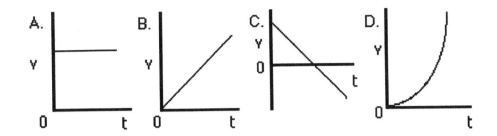

A. The slope of a v vs. t graph gives the acceleration. The slope of the line is zero, thus the object is moving with a constant velocity.

B. The slope of the line gives the acceleration. This graph shows the mass has a constant acceleration.

C. The slope also indicates there is constant acceleration. The acceleration has negative value. (Note the object has an initial velocity a velocity of zero and a negative final velocity.)

D. The velocity continues to increase, yet the rate is not constant.

6. If the velocity changes by the same amount for each similar time interval, what can you say about the acceleration ?

The acceleration would be constant. This means the velocity would continue to increase at equal intervals.

7. What are the units of acceleration in the SI system ?

meters per second squared - m/s^2 or m/s/s

8. When can the equation v $(v_f + v_o) / 2$ be used ?

When the acceleration is constant or uniform.

9. For a constant acceleration, how does the distance vary with time ?

The distance (x) varies as the square of the time (t^2). This means the accelerating body will **not** travel equal distances in equal periods of time. For bodies whose acceleration is positive, the distance traveled will increase during each interval of time.

10. What is the acceleration of an object in free fall ?

The acceleration due to gravity. ($g = 9.8$ m/s^2 or 32 ft/s^2)

Sample Quiz

(Remove the quiz from the book and test your knowledge of the chapter material as though you were taking an in-class quiz. Check you answers with the key at the book of the Study Guide.)

Completion

1. The time rate of change of position with the direction of change is
_____.

2. If the velocity, the average and instantaneous velocities are
_____ and the acceleration is _____.

3. When making a round-trip or returning to the starting point, the average velocity of the trip is _____ .

Multiple Choice

___ 4. Instantaneous speed
    A. is always constant
    B. depends on displacement
    C. is the magnitude of the instaneous velocity
    D. equals the average speed for constant acceleration

___ 5. The acceleration is constant
    A. when the instantaneous velocity is constant
    B. when the final velocity is less than the initial velocity
    C. when the velocity is zero
    D. none of the above

Problems

6. An object accelerates from rest at a rate of 2.0 m/s^2 .
   A. How far does the object travel during the 5.0 s interval ?
   B. What is the velocity of the object at the end of this time ?

7. A person drops a stone from the top of a building 64 ft tall.
   A. With what speed does the stone strike the ground ?
   B. How long does it take for the stone to reach the ground ?

8. A person throws a stone upward with a speed of 10 m/s from the top of a building which is 60 m above the ground.
   A. What is the maximum height of the mass above the ground ?
   B. With what velocity does the stone strike the ground ?
   C. How long is the ball in the air ?

Chapter 5   Motion in a Plane

Sample Problems

Components of Motion

Example   An object moves with a constant velocity of 20.0 m/s
northwest. How far north and west will the object travel in
30.0 s ?

Solution

$v_x = v(\cos \varnothing)$ $v_y = v (\cos \varnothing)$

$v_x = 20.0 (\cos 45)$ $v_y = 20.0 (\sin 45)$

$v_x = 14.1$ m/s west $v_y = 14.1$ m/s north

$x = v_x t$ $y = v_y t$

$x = (14.1)(30.0)$ $y = (14.1)(30.0)$

$x = 423$ m west $y = 423$ m north

Projectile Motion

Example 1   A ball is tossed horizontally at 10.0 m/s from the top of a
building 44.0 m high.
A. How long is the ball in the air ?
B. How far from the base of the building will the ball strike
the ground ?

Solution

given

$v_x = 10$ m/s $v_{oy} = 0$

$y = - 44$ m

$g = 9.80$ m/s^2

Note that x and y components of motion are independent - x has constant velocity and y has constant acceleration. First work with the y component of motion and find the time, then work with the x motion.

$$y = -(1/2) gt^2$$
$$-44 = -(1/2)(9.8)t^2$$
$$-44 = -4.9 t^2$$
$$t = 3.0 s$$
$$x = v_x t$$

$$x = (10.0)(3.0) = 3.0 \times 10^1 m$$

Example 2   A soccer ball is kicked from the ground at 20.0 m/s 53.0° above the horizontal.
A. How long does it take for the ball to strike the ground ?
B. How high will the ball move ?
C. What is the speed of the ball at its peak point ?
D. What is the acceleration of the ball at the peak point ?
E. How far from the point where the ball was kicked will the ball land ?
F. What is the speed of the ball when it strikes the ground?

Solution

given $v = 20.0$ m/s $\emptyset = 53.0°$
$v_x = v_o \cos\emptyset$ $v_{oy} = v_o \sin\emptyset$
$v_x = (20.0)(\cos 53.0)$ $v_{oy} = (20.0)(\sin 53.0)$
$v_x = 12.0$ m/s $v_{oy} = 16.0$ m/s

A. at the peak point $v = 0$, $g = 9.80$ m/s^2 , and $v_y = 16.0$ m/s

$$v = v_{oy} - gt$$
$$0 = 16.0 - (9.80)t$$
$$t = 1.63 s$$

B. y_{max} occurs at 1.63 s or when $v_y = 0$

$$y = v_{oy}t - (1/2)gt^2$$

$$y = (16.0)(1.63) - (1/2)(9.8)(1.63)^2$$
$$y = 26.1 - 13.0 m$$
$$y = 13.1 m$$

C. v_y (at the peak point = 0) and v_x which is constant is 12.0 m/s
Therefore the resultant is 12.0 m/s

D. The acceleration is in the y component of the motion. The acceleration is uniform throughout the time the object is in motion. The acceleration remains 9.8 m/s^2 **not** 0.

E. v_x = 12.0 m/s   t = ?   x = ?
$$x = v_x t$$

The time for the mass to reach the peak height is equivalent to the time for the mass to come back to the ground since it returns to the same height. Therefore the time the mass is in the air is 2 (1.63 s) = 3.26 s
$$x = (12.0)(3.26) = 39.2 \text{ m}$$

F. The speed of the mass is the same as it was when left the ground - 20.0 m/s

Centripetal Acceleration

Example 1   A car traveling at 20.0 m/s rounds a curve whose radius is 40.0 m. Find the centripetal acceleration of the car.

Solution
        v = 20.0 m/s   r = 40.0 m a_c = ?

$$a_c = v^2 / r$$
$$a_c = (20.0)^2 / 40.0 = 1.00 \times 10^1 \text{ m/s}^2$$

Example 2   An object traveling in a circle experiences a net acceleration of 3.0 m/s^2 which is directed toward the center of a circle whose radius is 3.0×10^2 m.
A. Calculate the speed of the object.
B. What is the period of the object?

A.　　$a_c = 3.0 \text{ m/s}^2$　　　　$r = 3.0 \times 10^2 \text{ m}$　　　$v = ?$

$$a_c = v^2 / r$$
$$v^2 = a_c r$$
$$v^2 = (3.0 \times 10^2)(3.0)$$
$$v = 3.0 \times 10^1 \text{ m/s}$$

B.　　$v = d / T$
$$T = d / v$$
$$T = (3.0 \times 10^2) / 3.0 \times 10^1 = 1.0 \times 10^1 \text{ s}$$

Solutions to Selected Problems from the Text

3. given : $v_x = 2.0 \text{ m/s}$; $v_y = 1.5 \text{ m/s}$; $t = 4.0 \text{ s}$
   treat the x and y components of the motion independently

   $x = v_x t$　　　　　　　　　　　$y = v_y t$
   $x = (2.0)(4.0)$　　　　　　　　　$y = (1.5)(4.0)$
   $x = 8.0 \text{ m}$　　　　　　　　　　$y = 6.0 \text{ m}$

7. given : v_x (river) = 4.0 ft/s ; v_y (boat) = 15 ft/s
   A. $t = 10 \text{ s}$
   　$x = v_x t$　　　　　　　　　　$y = v_y t$
   　$x = (4.0)(10)$　　　　　　　　$y = (15)(10)$
   　$x = 40 \text{ ft}$　　　　　　　　　$y = 150 \text{ ft}$

   B. $y = v_y t$
   　$t = v_y / y$
   　$t = 300 / 15 = 20 \text{ s}$

   　$x = v_x t$
   　$x = (4.0)(20) = 80 \text{ ft}$

10. given : $y = -30\,m$; $v_{yo} = -8.0\,m/s$; $g = 9.8\,m/s^2$

   (a) $v_f^2 = v_{yo}^2 - 2gy$

      $v_f^2 = (-8.0)^2 - 2(9.8)(-30)$

      $v_f = -26\,m/s$

   (b) $v_f = v_o - gt$

      $-26 = -8.0 - (9.8)(t)$

      $9.8\,t = -8.0 + 26$

      $9.8\,t = 18$

         $t = 1.8\,s$

15. given :

x motion	y motion
$v_x = 20\,m/s$	$v_{yo} = 0$
$t = ?$	$g = 9.8\,m/s^2$
$x =$	$t = ?$
	$y = -50\,m$

First work with the y - component of motion.

   $y = v_{oy} - (1/2)gt^2$

  $-50 = 0 \quad - (1/2)(9.8)t^2$

  $50 / 4.9 = t^2$

  $t = 3.2\,s$

  $x = v_x t = (20)(3.2) = 64\,m$

20. given : $v = 120\,ft/s$; $\emptyset = 45°$

   (a) $v_{yo} = v_o sin\emptyset$

      $v_{yo} = 120(sin45) = 84.9\,ft/s$

      at the peak height $v_y = 0$

      $v_y^2 = v_{yo}^2 - 2gy$

      $0^2 = (84.8)^2 - 2(32)y$

      $y = 113\,ft$

   (b) $R = v_o^2 sin2\emptyset / g$

      $R = (120)^2 sin90 / 32$

      $R = 450\,ft = 150\,yd$

25. given : $v_0 = 55.0$ m/s ; $\emptyset = 60°$; $t = 9.00$ s

   treat the x and y components of motion separately

   x motion   y motion

   $v_x = v_0 \cos\emptyset = (55.0) \cos 60 = 27.5$ m/s

   $v_y = v_0 \sin\emptyset = (55.0) \sin 60 = 47.6$ m/s

   $t = 9.00$ s $t = 9.00$ s

   $x = v_x t$ $y = v_0 t - (1/2)gt^2$

   $x = (27.5)(9.00)$ $y = (47.6)(9.00) - (1/2)(9.80)(9.00)^2$

   $x = 248$ m $y = 31.5$

30. first convert 60 mi/h into ft/s :

   (60 mi/h)(5280 ft/mi)(1 h /3600s) = 88 ft/s

   $a = v^2 / r$

   $25.8 = 88^2 / r$

   $r = 88^2 / 25.8$

   $r = 300$ ft ; therefore the diameter is 600 ft

34. $v = x / t$

   $x = 2\pi r = 2\pi (8.00 \times 10^5 m + 6.4 \times 10^6 m) = 7.2 \times 10^6$ m

   $t = 100$ min (60 s / 1 min) = 6.00×10^3 s

   $v = (7.2 \times 10^6) / (6.00 \times 10^3) = 1.2 \times 10^3$ m/s

   $a = v^2 / r$

   $a = (1.2 \times 10^3)^2 / (7.2 \times 10^6 m)$

   $a = 0.20$ m/s^2

Review Questions

1. What are "components of motion" ?

   The vector components in rectangular directions. By using the components, the motion may be analyzed in the particular (one-dimensional) direction and the net result is the vector sum of the component results.

2. Why are the signs or directions of the various quantities important in vertical projections ?

> To express the physical situation, the initial velocity being in the opposite direction to the acceleration.

3. For a vertical upward projection, what is the velocity and acceleration at the top of the path ?

> The velocity is instantaneously zero in the vertical component. The acceleration is constant - the acceleration due to gravity. Although the velocity is zero, gravity still acts. Therefore the acceleration does no equal zero.

4. What are the constant quantities for a horizontal projection ?

> The horizontal component of the velocity and the acceleration due to gravity.

5. What is the common factor between the components for a projection at an angle ?

> Time is common to both horizontal and vertical components of motion.

6. Explain why the range for a 30° and a 60° projection is the same for the same initial velocity.

> The range varies as sin 2Ø, and sin 2(30°) = sin 60° and sin 2(60°) = sin 120°.
> Since the sin120° = cos 30° and the cos 30° = sin 60°, they are equal. Check it out on a calculator !

7. For a projection angle of 45°, how may the range be increased ?

> $R_{max} = v_o^2 / g$, so by increasing the initial speed the range is increased. For example, if the speed were doubled, the range is increased by a factor of four.

8. Why is there necessarily an acceleration for uniform circular motion ?

Acceleration requires a change in velocity, and in uniform circular motion the velocity is continually changing (even though the speed is constant) in direction.

9. What would happen to an object in uniform circular motion if the centripetal acceleration went to zero ?

The object would fly off tangentially to the circular path since there would be no acceleration toward the center of the circle to cause it to remain in a circular path.

10. How is the centripetal acceleration of the object in uniform circular motion related to the object's period ?

The period (T) is the time for the object to make one complete orbit. Using $v = d / t$ or $v = d / T$, solve for T or $T = d / v$, where $d = 2\pi r$ (the circumference of a circle).   Then, $a_c = v^2 / r = (2\pi r / T)^2 / r = 4\pi^2 r / t^2$.

Sample Quiz

(Remove the quiz from the book and test your knowledge of the chapter materials as though you were taking an in-class quiz. Check your answers with the key at the end of the Study Guide.)

Completion

1. The velocity component of motion in the x-direction for a projection at an angle Ø relative to the x-axis is given by _____.

2. The _____ displacement is the same for a horizontal projection and an object dropped simultaneously from the same height.

3. A mass is moving clockwise in a circle. When the velocity is directed north the acceleration is directed _____.

Multiple Choice

____ 4. Which of the following quantities is constant for **all** types of projections ?
   A. horizontal velocity
   B. time of flight
   C. acceleration
   D. none of the above

____ 5. To decrease the centripetal acceleration of an object in uniform circular motion by one-half while maintaining the same orbit, the orbital speed would have to
   A. double
   B. decrease by one-half
   C. decrease by a factor of $1 / 2^{1/2}$
   D. decrease by one-fourth

____ 6. A ball is tossed with the same speed from the top of a cliff in the following ways :
   I. thrown horizontally from the cliff.
   II. thrown upward at a angle above the horizontal.
   III. thrown downward at an angle below the horizontal.
   The time the masses are in the air is greatest in which case ?
   A. I   B. II   C. III   D. all are the same

47

Problems

7. A ball is thrown horizontally from the top of a building 64 ft tall with a speed of 15 ft/s. What is the range of the ball ?

8. A golf ball is hit with a speed of 52 m/s 37° above the horizontal.
    A. How long is the ball in the air ?
    B. How high will the mass rise ?
    C. What is the range of the ball ?

9. A particle is in uniform circular motion with a centripetal acceleration of 25 m/s^2 in an orbit whose radius is 4.0 m.
    A. What is the speed of the particle ?
    B. What is the period of the particle ?

Chapter 6   Newton's Laws of Motion

Sample Problems

Newton's Second Law and Applications

Example 1   A 50.0 N net force acts horizontally on a an object whose
weight is 49 N.
A. Find the mass of the object.
B. Find the acceleration of the object.
C. If the mass starts from rest, find the speed of the mass
after 5.0 s.
D. How far does the mass travel during the 5.0 s ?

<u>Solution</u>

A. $w = mg$
$m = w / g$
$m = (49) / 9.8 = 5.0$ kg
B. $F = ma$
$a = F / m$
$a = 50.0 / 5.0 = 1.0 \times 10^1$ m/s^2
C. $v_f = v_o + at$

$v_f = 0 + (1.0 \times 10^1)(5.0) = 5.0 \times 10^1$ m/s

D. $d = v_o t + (1/2)\, at^2$

$d = (1/2)(1.0 \times 10^1)(5.0)^2 = 1.3 \times 10^2$ m

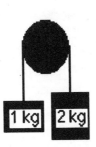

1 kg   2 kg

Example 2   Two masses are suspended from a pulley by a string as
shown.
A. Find the acceleration of the system.
B. Find the tension in the string.

<u>Solution</u>

For the right side, Newton's second law states :
$$m_2a = m_2g - T$$
For the left side, Newton's second law states :
$$m_1a = T - m_1g$$
Now add the two equations together.
$$m_1a + m_2a = m_2g - m_1g$$
$$(1) a + (2) a = 2 (9.8) - 1 (9.8)$$
$$3 a = 9.8$$
$$a = 3.3 \text{ m/s}^2$$

B. substitute into either of the equations for Newton's second law.
$$m_2a = m_2g - T$$
$$T = 2(9.8) - 2 (3.3) = 13 \text{ N}$$

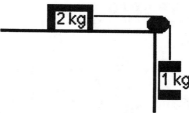

Example 3   In the figure above find the acceleration of the system if the surface is frictionless.

<u>Solution</u>

Hanging mass :	$m_1a = m_1g - T$
Mass on the surface :	$m_2 a = T$
add the two equations :	$m_1a + m_2a = m_1g$
	$(1) a + 2 (a) = 1 (9.8)$
	$3 a = 9.8$
	$a = 3.3 \text{ m/s}^2$

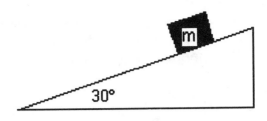

30°

Example 4   A mass slides down the frictionless surface as shown
above. Find the acceleration of the mass.

Solution

The component of the weight parallel to the surface tends
to pull the mass down the inclined plane. $F_x = mg \sin\emptyset$

Newton's law $ma = F_x$
$$ma = mg \sin\emptyset$$
$$a \ = (9.8) (\sin 30)$$
$$a \ = 4.9 \text{ m/s}^2$$
Note the acceleration is independent of the mass !!

Example 5   A 5.0 kg mass attached to the end of a string moves in a
uniform circle radius 0.50 m on the top of a horizontal
surface at 2.0 m/s. Find the tension in the string.

Solution

The tension in the string equals the centripetal force on
the mass.
$$F_c = T = mv^2 / r$$
$$T = (5.0)(2.0)^2 / 0.50$$
$$T = 4.0 \times 10^1 \text{ N}$$

Example 6   A car (mass 2.00×10^3 kg) travels around a curve (radius
50.0 m) with a speed of 3.00×10^1 m/s. Find the
minimum frictional force needed for the car to negotiate
the turn.

<u>Solution</u>

The force of friction (f) is what provided the centripetal force needed for the car to negotiate the turn.

$$f = F_c = mv^2 / r$$
$$f = (2.00 \times 10^3)(3.00 \times 10^1)^2 / 50.0$$
$$f = 3.60 \times 10^4 \text{ N}$$

Solution to Selected Problems from the Text

6. given : $m = 4.0$ kg ; $F = 10$ N ; $v_o = 3.0$ m/s ; $t = 2.0$ s

First solve the problem for the acceleration of the mass.
$$F = ma$$
$$a = F / m$$
$$a = 10 / 4.0$$
$$a = 2.5 \text{ m/s}^2$$
A constant net force produces a constant acceleration.
$$v_f = v_o + at$$
$$v_f = 3.0 + (2.5)(2.0) = 8.0 \text{ m/s}$$

10. given : $F = 4.0$ N ; $a = 0.25$ m/s^2
Work with this information to find the mass.
$$F_1 = ma_1$$
$$m = F_1/a_1 = 4.0 / 0.25 = 16 \text{ kg}$$

$$F_2 = ma_2 = (16)(1.5 \text{ m/s}^2) \qquad \text{(six times the original acceleration)}$$
$$F_2 = 24 \text{ N} \qquad\qquad\qquad \text{(six times the original force)}$$

16. This is also a two step problem. First find the acceleration of the system.

   let m_1 = 150 g and m_2 = 100 g

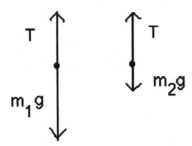

Newton's law for 150 g mass
$$m_1 a = m_1 g - T$$

Newton's law 100 g mass
$$m_2 a = T - m_2 g$$

add the two equations together :
$$m_1 a + m_2 a = m_1 g - m_2 g$$
$$(0.150 + 0.100)\, a = (0.150 - 0.100)\, 9.80$$
$$a = 2.0 \text{ m/s}^2$$

Constant forces produce constant acceleration :

$$x = v_0 t + (1/2)at^2$$
$$1.0 = (1/2)(2.0)(t^2)$$
$$t = 1.0 \text{ s}$$

20. Only work with the horizontal forces on the masses. The surface takes care of the weights of the bodies.

   (a) Write an equation for Newton's second law for the 2.0 kg mass.
   $$m_1 a = F - T \qquad \text{(T is the tension in the string connecting the masses)}$$
   Write an equation for Newton's second law for the 1.0 kg mass.

   $$m_2 a = T$$
   Now add the two equations together.
   $$m_1 a + m_2 a = F$$
   $$3 (a) = 18$$
   $$a = 6.0 \text{ m/s}^2$$

(b) The first or second equation may be used. The second is easier since it is shorter.

$$m_2 \, a = T$$
$$T = (1.0)(6.0) = 6.0 \text{ N}$$

25. (a) If m_3 were 2.0 kg, then the system would be at rest.

(b) If m_3 were 3.0 kg, then the 4.0 kg mass would move toward the left.

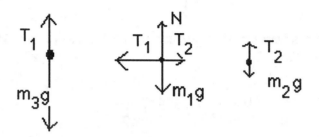

Write an equation for each body using Newton's laws of motion.

for m_3 $m_3 a = m_3 g - T_1$

for m_1 $m_1 a = T_1 - T_2$ (only the horizontal forces)

for m_2 $m_2 a = T_2 - m_2 g$

Now add the equations together. (T_1 and T_2 cancel)

$$m_1 a + m_2 a + m_3 \, a = m_3 g - m_2 g$$
$$9 \, a = (3 - 2)(9.8)$$
$$a = 1.1 \text{ m/s}^2 \text{ toward } m_3$$

(c) If m_3 were 1.0 kg the system would accelerate toward m_2.

The equations for Newton's laws would now be :

for m_3 $m_3 a = T_1 - m_3 g$ (since it now accelerates upward)

for m_1 $m_1 a = T_2 - T_1$

for m_2 $m_2 a = m_2 g - T_2$

combining the equations as in part B :

$$m_1 a + m_2 a + m_3 a = m_2 g - m_3 g$$
$$7 \, (a) = 2(9.8) - 1(9.8)$$
$$a = 1.4 \text{ m/s}^2 \text{ toward } m_2$$

To find the tension in the strings, substitute back into the 1st and 3rd equations

$$m_3 a = T_1 - m_3 g \qquad\qquad m_2 a = m_2 g - T_2$$
$$(1.0)(1.4) = T_1 - (1.0)(9.8) \qquad (2.0)(1.4) = (2.0)(9.8) - T_2$$
$$T_1 = 11\ N \qquad\qquad T_2 = 17\ N$$

30. given : $m = 2.0$ kg ; $v_0 = 8.0$ m/s ; $t = 2.0$ s

    A. $F = + 10\ N$

       $F = ma$

       $a = F/m = 10\ N\ /\ 2.0\ kg = 5.0\ m/s^2$

       $v_f = v_0 + at$

       $v_f = (8.0) + (5.0)(2.0) = 18$ m/s in the +x direction

    B. $F = -10\ N$; therefore $a = -5.0\ m/s^2$

       $v_f = (8.0) + (-5.0)(2.0) = 2.0$ m/s in the - x direction.

35. First find the acceleration of the mass.

    given : $v_0 = 25$ m/s ; $v_f = 0$; $x = 100$ m ; find a

$$v_f^2 = v_0^2 + 2ax$$
$$0^2 = 25^2 + 2a(100)$$
$$a = -3.1\ m/s^2$$

Now find the force. $F = ma$

    $F = (5.0)(3.1) = 15.5\ N$

The frictional force must be greater than or equal to 15.5 N for the mass to remain stationary. The force between t he box and the truck is only 15 N. Therefore the box would move relative to the truck.

40. First find the speed of the moon.

    $v = x / t$

    $x = 2 \pi r$ (where r = mean distance between the Earth and the moon)

    $x = 2 \pi (3.8 \times 10^8\ m) = 2.4 \times 10^9\ m$

    $t = 29.5$ days (24 h/1 day) (3600 s / 1 h) $= 2.5 \times 10^6$ s

    $v = (2.4 \times 10^9)\ /\ (2.5 \times 10^6) = 9.6 \times 10^2$ m/s

(a) $F = mv^2/r$

$F = (7.4 \times 10^{22} \text{ kg})(9.6 \times 10^2)^2 / 2.4 \times 10^9$

$F = 2.8 \times 10^{19} \text{ N}$

(b) The force is equal and opposite (Newton's third law of motion)

$F = 2.8 \times 10^{19} \text{ N}$

Review Questions

1. Why is Newton's first law called the the law of inertia ?

Inertia, the property of matter that resists changes in motion, describes the effects of Newton's first law, an object remains in motion or stationary (that is resists changes in motion) unless acted upon by a force that produces changes in motion.

2. Who first formulated Newton's first law of motion ?

This is attributed to Galileo.

3. What are the cause and effect of Newton's second law ?

The cause is force and the affect is acceleration or motion.

4. Explain how Newton's first law is given by Newton's second law.

If the net force is zero, then there is no acceleration ($F = ma$) and the velocity is constant (or zero), which is what the first law states.

5. Give the standard combination for the units of newton and pound.

$F = ma$, and kg-m/s^2 = N and slug-ft/s^2 = lb.

6. Why do we say that the equivalent weight of 1 kg is 2.2 lb ?

> A kilogram is a unit of mass and not weight or force. In a uniform gravitational field with $g = 9.8 \text{ m/s}^2 = 32 \text{ ft/s}^2$, then the weight force of 1 kg is equivalent to 2.2 lb.

7. What supplies the centripetal acceleration for uniform circular motion ?

> A centripetal force as suppled by a physical pull on a string or gravity, which may be described in terms of mass and motional parameter as $F = ma = mv^2/r$.

8. What is the essence of Newton's third law ?

> That forces occur in equal and opposite pairs.

9. What is action and reaction ?

> Terms used to describe the force pair of Newton's third law. Either force may be the action or reaction.

10. If forces occur in equal and opposite pairs in accordance with Newton's third law, how can there be a net force since they would cancel out ?

> The force pair of the third law acts on different bodies. The net force of the second law is concerned with the forces acting on a particular body.

Sample Quiz

(Remove the quiz from the book and test you knowledge of the chapter materials as though you were taking an in-class quiz. Check your answer with the key at the back of the Study Guide.)

Completion

1. The inertia of a body is directly proportional to its _____.

2. The acceleration of an object acted on by a net force is inversely proportional to its _____.

3. A reaction force acts on the body producing or exerting the

   _____.

Multiple Choice

____ 4. According to Newton's second law, which of the following is a unit of mass ?

   A. $N\text{-}s^2/m$ B. $N\text{-}s$ C. $m^2\text{-}N$ D. $s^2\text{-}m/N$

____ 5. A centripetal force is described by
   A. Newton's first law
   B. Newton's second law
   C. Newton's third law
   D. all of the above

Problems

6. Two forces 6.0 N and 4.0 N, act in opposite direction on a mass of 0.50 kg resting on a frictionless surface.
   A. What is the net acceleration of the mass ?
   B. How far will the mass move in 3.0 s ?

7. A 2.0 kg mass moves in a uniform circle has its speed increased from 1.5 m/s to 4.5 m/s while maintaining the same orbit. By how much must the force on the mass be increased to do this ?

8. In the figure above F = 60.0 N, find the acceleration of the system and the tension in the cable connecting the masses.

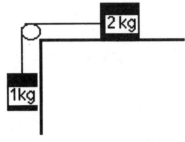

9. The system above is at rest. Find the force needed to keep the system motionless.

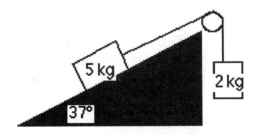

10. A 5.0 kg mass moves down a frictionless 37 ° inclined plane. Attached to the mass is a string which is connected to a 2.0 kg mass. Find the acceleration of the system.

Chapter 7   Gravity and Friction

Sample Problems

Newton's Law of Gravitation

Example 1   Two persons each of 70.0 kg stand 2.00 m apart. Find the force of gravitational attraction between the two masses.

Solution

$$F = G \, m_1 \, m_2 / r^2$$
$$F = (6.67 \times 10^{-11}) \, (70)(70) / 2.00^2$$
$$F = 8.17 \times 10^{-8} \, N$$

Example 2   Find the gravitational force of attraction between the sun and Jupiter. The average distance between Jupiter and the sun is 7.78×10^{11} m and the mass of Jupiter is 1.9×10^{27} kg and the sun is 2.0×10^{30} kg.

Solution

$$F = G m_1 m_2 / r^2$$
$$F = (6.67 \times 10^{-11})(2.0 \times 10^{30}) \, (1.9 \times 10^{27}) / (7.78 \times 10^{11} \, m)^2$$
$$F = 4.19 \times 10^{23} \, N$$

Example 3   The mass of the moon is 7.4×10^{22} kg and the earth is 6.0×10^{24} kg. The mean distance from the two bodies is 3.84×10^8 m. From this information find the
A. speed of the moon as it moves around the earth.
B. the period of the moon based on this information.

The gravitational force provides the centripetal force in this situation.

<u>Solution</u>

$$F = G m_1 m_2 / r^2$$
$$F = (6.67 \times 10^{-11}) (7.4 \times 10^{22})(6.0 \times 10^{24}) / (3.84 \times 10^8)^2$$
$$F = 2.0 \times 10^{20} \text{ N}$$

A Closer Look at g and Apparent Weightlessness

Example   A 70.0 kg person stands on a set of scales as he moves in
an elevator. What is the person's weigh on the scales
when the elevator is
A. at rest ?
B. moving upward at a constant sped of 2.0 m/s ?
   downward at 2.0 m/s ?
C. accelerating upward at 2.0 m/s^2.
D. accelerating downward at 2.0 m/s^2

<u>Solution</u>

$$ma = F_s - mg$$
$$F_s = ma + mg$$

A. $a = 0$ therefore $F_s = mg = (70.0)(9.8) = 686$ N

B. since $a = 0$, the weight is the same as in part A.

C. $ma = F_s - mg$ (since the net force is upward)
   $F_s = ma + mg = (70)(2.0 + 9.8) = 826$ N

D. $ma = mg - F_s$ (since the net force is downward)
   $F_s = mg - ma = (70.0)(9.8 - 2.0) = 546$ N

Coefficients of Friction

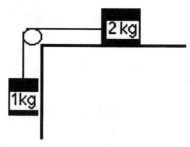

Example 1   In the figure above, the system is at rest.
A. Find the value of the frictional force needed to keep the system at rest.
B. Find the value of the coefficient of static friction needed to keep the system at rest.

<u>Solution</u>

A. Applying Newton's second law to the hanging mass :

$m_1 a = m_1 g - T$

$T = m_1 g = (1.0)(9.8) = 9.8 \text{ N}$ (since a = 0)

Applying Newton's second law to the mass on the surface :

$m_2 a = T - f$

$T = f = 9.8 \text{ N}$ (since a = 0)

B. $\mu = f / N$

$\mu = f / N$ (since N = $m_2 g$)

$\mu = 9.8 / 19.6 = 0.5$

Example 2   In the example above suppose a slight push were given to the system in order to start the system in motion.   Find the value for the acceleration if $\mu_k = 0.20$.

<u>Solution</u>

from the problems above $m_1 a = m_1 g - T$

$m_2 a = T - \mu_k m_1 g$

add the two equations

$(m_1 + m_2)a = m_1 g - \mu_k m_2 g$

$(3) a = 9.8 - (0.2)(2.0)(9.8)$

$a = 2.0 \text{ m/s}^2$

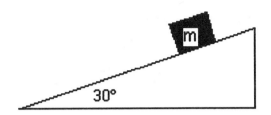

Example 3   The mass in the diagram above accelerates down the
plane with an acceleration of 1.0 m/s^2. Find the value of
the coefficient of kinetic friction.

Solution

Apply Newton's 2nd law of motion for the axis on the
surface.

$ma = mgsin\emptyset - f$ $\mu = f / N$
                                $N = mgcos\emptyset$
                                $f = \mu mgcos\emptyset$
                                $ma = mgsin\emptyset - \mu mgcos\emptyset$

(note the mass cancels)
$1.0 = (9.8)sin30 - \mu (9.8)(cos 30)$
$1.0 = 4.9 - 8.5 \mu$
$8.5 \mu = 3.9$
$\mu = 0.46$

Solutions to Selected Problems from the Text

5.   given : $F = 2.03 \times 10^{-5}$ N ; $r = 3.30$ m

$F = G m_1 m_2 / r^2$

$2.03 \times 10^{-5} = (6.67 \times 10^{-11}) m^2 / 3.30^2$

$2.03 \times 10^{-5} = 6.12 \times 10^{-12} m^2$

$m = 1.8 \times 10^3$ kg

6.   given $k = 250$ N/m ; $m = 5.0$ kg
The weight of the mass stretches the spring.
$F = mg = (5.0)(9.8) = 49$ N

$F = kx$
$x = F / k$
$x = 49 / 250 = 0.20$ m

10.   given : $r = 5 R_e$ (4 radii above the Earth plus the radius of the Earth)

$g = G m / r^2$

$g = (6.67 \times 10^{-11})(6.0 \times 10^{24}) / (3.2 \times 10^7)^2$

$g = 0.39 \text{ m/s}^2$

alternate solution - proportion

$g_1 / g_2 = r_2^2 / r_1^2$

$9.8 / g_2 = (5R_e)^2 / R_e^2$

$g_2 = 0.39 \text{ m/s}^2$

15. (a) $g = Gm / r^2$

$g = (6.67 \times 10^{-11})(6.0 \times 10^{24}) / (8.0 \times 10^6)^2$

$g = 6.3 \text{ m/s}^2$

(b) The acceleration at this point only depends upon the mass of the Earth not the body. Thus the acceleration for the two masses would be the same.

22.   There are several ways to compute the mass. The Earth travels around the Sun because of gravitational forces. The mean distance from the Earth to the Sun is 1.5×10^{11} m and the time it takes for the Earth to revolve around the Sun is 365 days = 3.15×10^7 s.

$v = x / t$

$v = 2 \pi r / t$

$v = 2\pi (1.5 \times 10^{11} \text{m}) / (3.15 \times 10^7)$

$v = 3.0 \times 10^4 \text{ m/s}$

$F = m_E v^2 / r$

$F = G m_E m_S / r^2$

Set the two forces equal :

$$m_E v^2 / r = G\, m_E m_S / r^2 \qquad\qquad \text{note } m_E \text{ cancels}$$

$$(3.0 \times 10^4)^2 / 1.5 \times 10^{11} = (6.67 \times 10^{-11})\, m_S / (1.5 \times 10^{11})^2$$

$$6.0 \times 10^{-3} \quad = 3.0 \times 10^{-33}\, m_S$$

$$m_S = 2.0 \times 10^{30}\ kg$$

26. $ma = mg + mg$
  $ma = 2\,mg$
   $a = 2g$
   $a = 2(9.8) = 19.6\ m/s^2$

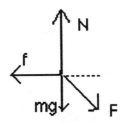

32.

Look at the horizontal forces first : $ma = F\cos 60 - f$
$$f = \mu N$$
Combine the two equations $\qquad ma = F\cos 60 - \mu N$
Look at the vertical forces for N $\qquad 0 = N - F\sin 60 - mg$
Substituting $\qquad\qquad ma = F\cos 60 - \mu_k mg - \mu_k F\sin 60$

$$(2.0)(a) = (40)(\cos 60) - (0.27)(2.0)(9.8) - (0.27)(40)(\sin 60)$$
$$(2.0)(a) = 20 - 5.3 - 9.4$$
$$a = 2.7\ m/s^2$$

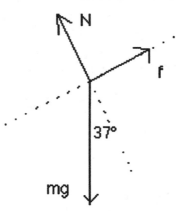

35.

As you solve the problem it is much easier to rotate the x and y axis. Make the surface the x axis.

(a) $F_x = mg \sin37$ (force tending to make block move down)

$F_x = (250)(9.8)(\sin37) = 1.5 \times 10^3$ N
Now find the force of static friction $f = \mu N$

The force N is in the y component. $0 = N - mg\cos37$

$N = (250)(9.8)(\cos37) = 1.96 \times 10^3$ N
The force of friction is $f = \mu N = (0.50)1.96 \times 10^3$ N $= 980$ N

Since the force of friction is less than F_x the object will slide down the plane.

(b) If the mass is not to move, the net force must be zero.
A force must be applied up the plane to keep the mass stationary.
$F = 1.5 \times 10^3$ N $- 9.8 \times 10^2$ N $= 5.2 \times 10^2$ N up the plane.

(c) Kinetic friction less than static friction since μ is different.
Calculate the frictional force first. $f = (0.30)(1.96 \times 10^3) =$
$f = 5.9 \times 10^2$ N
More force is needed up the plane since friction is less.
$F = 1.5 \times 10^3$ N $- 5.9 \times 10^2$ N $= 9.1 \times 10^2$ N up the plane

40. The free body diagram is similar to problem 35. Remember to rotate the axis since it is an inclined plane.

x forces : $\quad ma = mg \sin30 - f$

$\qquad\qquad ma = mg \sin30 - \mu_k N \qquad\qquad$ (since $f = \mu_k N$)

y forces : $\quad 0 = N - mg \cos30$

substitute $\quad ma = mg\sin30 - \mu_k mg\cos30$

$\qquad\qquad a = (9.8)(\sin30) - 0.25(9.8)(\cos30)$ (note m cancels)

$\qquad\qquad a = 2.8 \text{ m/s}^2$

Since constant forces produce constant acceleration :

$x = v_o t + (1/2)at^2$

$3.0 = 0 + (1/2)(2.8)(t^2)$

$t = 1.5 \text{ s}$

44. To negotiate the turn the static frictional force should be greater than or equal to the centripetal force.

$\qquad F_c = mv^2 / r$

$\qquad f = \mu_s N$

Since the body is on a horizontal surface $N = mg$

$\qquad f = \mu_s \, mg$

$\qquad \mu_s mg \geq mv^2 / r \qquad\qquad$ Note the mass cancels

Compare the left side of the equation with the right side.

$\qquad \mu_s g = (0.80)(9.8) = 7.84 \text{ m/s}^2$

$\qquad v^2 / r = 25^2 / 75 = 8.33 \qquad\qquad$ 90 km/h = 25 m/s

Since the frictional force is less than the centripetal force, the car would slide outward.

Review Questions

1. Distinguish between "big" G and "little" g.

    "Big" G or simple G is the universal gravitational constant which is the constant proportionality in Newton's law of gravitation. "Little" g or the acceleration due to gravity depends on the gravitational force, and $g = GM / R^2$.

2. What masses are the m's in Newton's law of gravitation ?

    The m's can be any particles. Every particle is attracted toward every other particle. For real particle forces, the net force is the vector sum of the individual sum of the individual particle forces, which is difficult to compute. Uniform spherical objects may be treated as though all of the mass is concentrated at the center or center of mass.

3. What is measured or balanced by a spring scale ?

    The weight force of the object being measured is balanced by the spring force, $F = kx$.

4. Approximately what percent of its value on Earth is the acceleration due to gravity at an altitude of 100 mi (160 km) ?

    About 95 %.

5. How does the speed of satellites vary with altitudes and what is a synchronous satellite ?

    The orbital speed as $1 / (R_e + h)^{1/2}$ with altitude, that is the speed decreases with altitude. A synchronous satellite is one at an altitude such that its orbital period is the same as the rotational period of Earth, and the satellite is relatively stationary over one spot.

6. Is it ever possible to have true weightlessness ?

Only if the vector sum of two or more gravitational forces is zero. Near the Earth, the force of gravity is always nonzero so objects have weight by definition.

7. What is friction ?

The ever present resistance to motion between contacting materials or with a medium.

8. What is the greatest source of friction between metal surfaces ?

Local or contact welding at asperating junctions.

9. How is the classical law that friction is directly proportional to the load expressed mathematically ?

The load is expressed in terms of the normal reaction force (N) of the surface acting on an object so that components of load are not a problem, and $f = \mu N$, where μ is the coefficient of friction.

10. What is terminal velocity and how does it differ for objects of the same shape, but have different weights ?

Terminal velocity is the constant velocity of a falling object when the resistive force of air friction is equal to the object's weight. Heavier objects reach terminal velocities later than similar lighter objects.

Sample Quiz

(Remove the quiz from the book and test your knowledge of the chapter material as though you were taking an in-class quiz. Check your answers with the key at the back of the Study Guide.)

Completion

1. From the law of gravitation, the SI units of the universal gravitational constant are _____.

2. An object with twice the mass of another has twice the force of gravity in free fall, but falls at the same rate because it has_____.

3. According to the classical laws of friction, the frictional force is independent of _____ and _____.

Multiple Choice. Choose the best answer.

___4. An Earth satellite
   A. experiences zero g
   B. is synchronous at any altitude
   C. has an orbital speed for a given altitude
   D. has a specific orbital speed for a given altitude

___5. The force of friction that varies with the applied force is
   A. static friction
   B. kinetic friction
   C. sliding friction
   D. all of the above

___6. Which of the following factors does not affect the terminal speed ?
   A. its mass
   B. its volume
   C. initial velocity
   D. air resistance

Problems

7. A small mass is in the vicinity of a large 100 kg sphere and is 15 m from its center. Find the initial acceleration of the small mass.

8. A 6.0 kg block is initially at rest on a 30° inclined plane. If the coefficients of static and kinetic friction between the block and the plane are 0.65 and 0.40 respectively, what happens to the block ?

9. A 2.0 kilogram block, initially moving at 10 m/s slides on a rough surface. If the coefficient of kinetic friction between the surfaces is 0.25, find
    A. the deceleration of the mass.
    B. the distance the mass moves before it stops.

Chapter 8 Work, Energy, and Power

Sample Problems

Work

Example 1   A 3.0 kg mass is lifted a height of 2.0 m. Find the work
done against gravity.

Solution

$m = 3.0\,\text{kg}\,;\,d = 2.0\,\text{m}$
first find the force : $F = mg = (3.0\,)(9.8) = 29.4\,\text{N}$

$W = Fd$
$W = (29.4\,\text{N})(2.0\,\text{m}) = 59\,\text{J}$

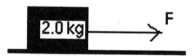

Example 2   A 2.0 kg mass is pulled 10.0 m with a force $F = 10\,\text{N}$ as
shown above. The coefficient of kinetic friction is 0.20.
A. Calculate the work done by the force **F**.
B. Calculate the work done by the frictional force.
C. Calculate the net work done.

Solution

given $m = 2.0\,\text{kg}\,;\,F = 10\,\text{N}\,;\,d = 10\,\text{m}$

A. $W = Fd$

$W = (10)\,(10) = 1.0 \times 10^2\,\text{J}$

B.   find the frictional force
$f = \mu_k N$
$f = (0.20)(2.0)(9.8) = 3.9\,\text{N}$ (since $N = mg$)
$W = -(3.9)(10)$
$W = -39\,\text{J}$

C. $W_{net} = W_F + W_f + W_{mg} + W_N$
    $W_{net} = 100 \text{ J} + -39 \text{ J} + 0 + 0 = 61 \text{ J}$

Example 3   Find the work required to compress a spring (k = 100 N/m) a distance of 20.0 cm.

Solution

$W = (1/2)kx^2$
$W = (1/2)(100)(0.20)^2$
$W = 2.0 \text{ J}$

Energy

Example 1   Calculate the change in kinetic energy for Example 2 in the preceding group.

Solution

$W = \Delta K = 61 \text{ J}$
The problem can be worked using kinematics and Newton's laws.

$ma = F - f$
$ma = 10 - 3.9$
$ma = 6.1$
$a = 3.05 \text{ m/s}^2$

$v^2 = v_o^2 + 2ax$
$v^2 = 0^2 + 2(3.05)(10)$
$v = 7.8 \text{ m/s}$
$\Delta K = K_f - K_o$
$\Delta K = (1/2)mv^2 - (1/2)mv_o^2$
$\Delta K = (1/2)(2.0)(7.8)^2 - 0$
$\Delta K = 61 \text{ J}$

Example 2   A 2.0 kg box is lifted from a table 2.0 m from the floor to a shelf 3.0 m from the floor.
            A. Calculate the work done against gravity.
            B. Calculate the increase in potential energy.

$$m = 2.0 \text{ kg} \quad ; \quad d = 1.0 = \Delta h$$

A. $W = Fd$
$W = mgd$
$W = (2.0)(9.8)(1.0) = 2.0 \times 10^1 \text{ J}$

B. $PE = mg\Delta h$
$PE = (2.0)(9.8)(1.0) = 2.0 \times 10^1 \text{ J}$

Conservation of Energy

Example 1 A 2.0 kg stone is dropped from a cliff whose height is 100 m above a valley.
A. Calculate the initial potential energy of the stone with respect to the valley.
B. Calculate the potential and kinetic energies of the mass at a height of 40 m. What is the speed of the mass at this height ?
C. Calculate the potential and kinetic energy 3.0 s after it is released.
D. Find the speed of the stone as it strikes the ground.
E. How could the stone strike the ground with a greater speed ?

Solution

$$m = 2.0 \text{ kg} ; \ h = 100 \text{ m from the ground}$$

A. $PE_{80} = mgh$
$PE_{80} = (2.0)(9.8)(100) = 1960 \text{ J}$

B. $PE_{40} = (2.0)(9.8)(40) = 784 \text{ J}$
Since the total energy is constant the kinetic energy equals :
$1960 \text{ J} - 784 \text{ J} = 1176 \text{ J}$
$K = (1/2)mv^2$
$1176 = (1/2)(2.0)v^2$
$v = 34 \text{ m/s}$

C. First find either the height or the velocity using constant acceleration.

$$v = v_o - gt$$
$$v = 0 - (9.8)(3.0) = -29.4 \text{ m/s}$$

$$K = (1/2)mv^2$$
$$K = (1/2)(2.0)(-29.4)^2 = 864 \text{ J}$$
$$PE = 1960 \text{ J} - 864 \text{ J} = 1096 \text{ J}$$

D. All of the mechanical energy is in the form of K. K = 1960 J
$$1960 = (1/2)(2.0)v^2$$
$$v = 44.3 \text{ m/s}$$

E. If the stone were given an initial speed, the mass would have a greater speed when it strikes the ground.

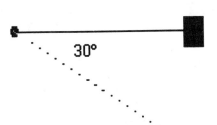

Example 2   A pendulum has a length of 1.0 m. Attached to the end is a 0.50 kg mass. The mass is released when the cord is in a horizontal position.
A. Find the maximum speed of the mass.
B. Find the speed of the mass when the angle Ø is 30° below the horizontal.

Solution

Use the lowest point of the swing as the 0 gravitational potential energy level.

A. $mgh_{horizontal} = (1/2)mv^2_{bottom}$

$(9.8)(1.0) = (1/2)v^2$
$v = 4.4\ m/s$

B. $mg\Delta h_{at\ 30°} + (1/2)\ mv^2 = mgh_{horizontal}$

$g\ (L - Lcos60) + (1/2)v^2 = gL$

$g\ (1/2)L + (1/2)v^2 = gL$
$(1/2)v^2 = (1/2)gL$

$v^2 = (9.8)(0.5)$
$v = 2.2\ m/s$

Power

Example 1   A motor can lift a mass of 2.5 kg a height of 10.0 m in 5.0 s. Find the power of the motor in watts and in horsepower.

Solution

$m = 2.5\ kg\ ;\ h = 10.0\ m\ ;\ t = 5.0\ s$

$F = mg = (2.5\ )(\ 9.8) = 24.5\ N$
$v = x/t$
$v = 10/5 = 2.0\ m/s$
$P = F\ v$
$P = (24.5)(2.0) = 49\ W$
$P = (49\ W)\ (1\ hp\ /\ 746\ W) = 6.6 \times 10^{-2}\ hp$

Example 2   A car engine requires a force of 200 lb to keep a car moving against the air and friction with a constant speed of 50 mi/h.   Find the power the motor must exert.

Solution

$(50\ mi/h\ )(\ 5280\ ft\ /\ 1\ mi)\ (1\ h\ /\ 3600\ s) = 73\ ft/s$

$P = F\ v$
$P = (200\ lb)(73\ ft/s) = (1.46 \times 10^4\ ft\text{-}lb\ /\ s)$

76

$$P = (1.46 \times 10^3 \text{ ft-lb/s})(1 \text{ hp} / 550 \text{ ft-lb/s}) = 26 \text{ hp}$$

Solutions to Selected problems from the Text

5. given   : $W = 2.0 \times 10^3 \text{ J}$; $d = 5.0 \text{ m}$
$$W = F d$$
$$F = W / d$$
$$F = (2.0 \times 10^3) / 5.0$$
$$F = 4.0 \times 10^2 \text{ N}$$

9. given   : $m = 10 \text{ kg}$; since v constant then $a = 0$; $d = 4.0 \text{ m}$

   A. $f = \mu_k N$
   $$f = \mu_k \, mg \qquad \text{(since } N = mg)$$
   $$f = (0.50)(10)(9.8) = 49 \text{ N}$$

   $$W = f \, d \cos 180$$
   $$W = (49)(4.0)(-1) = -2.0 \times 10^2 \text{ J}$$

   B.   by Newton's 2nd law : $0 = F - f$
   therefore $F = 49 \text{ N}$

   C.   calculate the work done by the applied force
   $$W = Fd$$
   $$W = (49)(4.0) = 2.0 \times 10^2 \text{ J}$$
   Since the net force is zero the net work is 0
   $$W_{net} = F_{net} \, d$$

15. given   : $m = 1.00 \times 10^3 \text{ kg}$; $v = 90 \text{ km/h} = 25 \text{ m/s}$

$$W = \Delta K$$
$K_o = 0$ since the initial velocity is 0
$$K_f = (1/2)mv_f^2$$
$$K_f = (1/2)(1.0 \times 10^3)(25)^2$$

$$K_f = 3.1 \times 10^5 \text{ K}$$
$$W = \Delta K = +3.1 \times 10^5 \text{ J}$$

21. given: $PE = 135 \text{ J}$; $k = 3.0 \times 10^3 \text{ N/m}$

$$PE = (1/2)kx^2$$
$$135 = (1/2)(3.0 \times 10^3) x^2$$
$$135 = (1.5 \times 10^3) x^2$$
$$x = 0.30 \text{ m}$$

24. given : $v_{o1} = 30 \text{ mi/h}$; $v_{o2} = 60 \text{ mi/h}$

The net work done equals the change in kinetic energy.
In both cases the final kinetic energy equals zero.

$$W = \Delta K = K_f - K_o = 0 - K_o$$
$$W = F d$$

$$\frac{F_1 d_1}{F_2 d_2} = \frac{\Delta K_1}{\Delta K_2}$$

note that since the force is constant that it cancels in the equation.

Therefore a v^2 - d relationships exists

the same as for constant acceleration.

Since the velocity is doubled the distance would be quadrupled (x 4).

28. given : $h = 200 \text{ m}$
$$PE_{top} = mgh$$
At the new location the kinetic energy will be twice the potential energy.
$$PE_{top} = PE_{new} + K_{new}$$
$$PE_{top} = PE_{new} + 2PE_{new}$$
$$PE_{top} = 3PE_{new}$$
$$mgh_{top} = 3mgh_{new}$$

$$h_{top} = 3\,h_{new}$$
$$200\,/\,3 = h_{new} = 66.7 \text{ m}$$

33. given : mass = 1.0 kg   ; v_0 at the top is 0

A. $mgh_{top} = PE_A + K_A$

$mgh_{top} = mgh_A + (1/2)mv^2_A$

$(1.0)(9.8)(10) = (1.0)(9.8)(2.0) + (1/2)(1.0)(v_A^2)$

$98 \text{ J} = 19.6 \text{ J} + 0.5\,v_A^2$

$v_A = 12.5 \text{ m/s}$

Use a similar set up for point B.

$98 \text{ J} = (1.0)(9.8)(6.0) + (1/2)(1.0)(v_B^2)$

$39.2 = (0.5)v_B^2$

$v_B = 8.9 \text{ m/s}$

B. $W = \Delta K$

$f\,d = -98 \text{ J}$

$f = \mu_K mg$

$\mu_K mgd = -98$

$(0.50)(1.0)(9.8)(-1)d = -98$

$d = 20 \text{ m}$

C. From zero to point A, the work is done by gravity. From A to B the work done is against gravity. From B to C the work is done by gravity.

38. given : $v_0 = 4.0$ m/s ; $v_f = 6.0$ m/s ; F = 30 N

$P = Fv$

$P = F\,\Delta v$

$P = (30)\,(6.0 - 4.0)$

$P = 60 \text{ W}$

44. given : $h = 25$ ft ; $R = 100$ ft^3/min ; $D = 62.4$ lb/ft^3

First find the weight of water lifted per unit time
$(100$ ft^3/min) $(62.4$ lb/ft$^3) = 6240$ lb / min $= 104$ lb/s
$P = F v$
$P = mg$ (y/t)
$P = (104$ lb/s)$(25$ ft) $= (2.6 \times 10^3$ ft-lb/s)$(1$ hp / 550 ft-lb/s)
$P = 4.7$ hp

Review Questions

1. What is required for work ?

In general, motion, i.e., a force acting through a distance, which produces motion.

2. What does negative work mean ?

That the force and the displacement are in the opposite directions, in bringing a moving object to rest.

3. What is the work on a graph of F vs. x for a spring ?

The work is equal to the area under the curve or straight line (since $F = kx$). This is the area of a triangle and
$A = W = (1/2)$ kx^2.

4. If v_f and v_o are the final and initial velocity does of an object, is the change in kinetic energy proportional to $(v_f - v_o)^2$?   Explain.

NO. The change in kinetic energy is $\Delta K = K_f - K_o$
$\Delta K = (1/2)mv_f^2 - (1/2)mv_o^2 = (1/2)$ m $(v_f^2 - v_o^2)$. The difference of the squares is not the same as the square of the differences.
(Example : $(3 - 2)^2 \neq 3^2 - 2^2$)

5. Is there an absolute zero reference for gravitational potential energy in the form mgh ?

> No, the reference is arbitrary. The <u>change</u> in potential energy is usually the important consideration and the Δh does not depend in the reference.

6. What is meant by rest energy ?

> An object has energy by virtue of its rest mass. In some instances, this mass is converted into energy according to the equation $E_0 = m_0 c^2$, and we say that mass is a form of energy .

7. Distinguish between the conservation of total energy and the conservation of mechanical energy .

> The total energy is conserved in every physical situation. That is, the energy is somewhere in some form. If the mechanical energy is conserved, the sum of the kinetic and potential energies is constant. This generally an ideal case.

8. Why are mechanical systems generally non-conservative ?

> Due to frictional losses.

9. How can one increase power ?

> Since power is the work per unit time, the power could be increased by adding more work in the same amount of time or doing the same work in less time.

10. What does the efficiency of a machine tell us ?

> It tells what you get out for what you put in - how much useful work is done or how much energy is lost.

Sample Quiz

(Remove the quiz from the book and test your knowledge of the chapter material as though you were taking an in-class quiz. Check your answers with the key at the back of the Study Guide.)

Completion

1. The work doe by a force acting perpendicular to the direction of motion is _____.

2. Work is equal to the change in kinetic energy. This statement is called the _____.

3. The efficiency of a machine is always _____.

Multiple Choice.

____ 4. If the kinetic energy of an object decreases, then
   A. the potential energy must increase
   B. the work is done on the object.
   C. the mechanical energy is conserved.
   D. none of the above

____ 5. Which of the following is a true statement about kinetic energy ?
   A. Kinetic energy is usually positive but can sometimes be negative.
   B. Kinetic energy must always be greater than zero.
   C. Kinetic energy may be either 0 or a positive number.
   D. Kinetic energy must always be equal to the potential energy.

____ 6. The unit of hp-h (horse power - hour) would be a unit of
   A. power   B. energy   C. speed   D. rest mass

Problems

7. A 2.0 kg mass is dropped from the top of a 30 m building.
   A. How muck work is done by gravity when the mass falls 10 m ?
   B. What is the speed of the mass at this time ?

8. A machine has an efficiency of 75%. If its power input is 6.0 hp, how much work could the machine do in 10 minutes ?

9. A 10 N net force is applied to a 5.0 kg mass, initially at rest, for 5.0 s.
   A. How much work was done on the mass.
   B. Find the change in kinetic energy of the mass.
   C. Find the speed of the mass after the acceleration period.

Chapter 9   Momentum

Sample Problems

Example 1   A net force of 5.0 N acts on a 2.0 kg mass for 10.0 s.
                 A. Find the change in momentum of the mass.
                 B. Find the change in velocity for the mass.

Solution

     A. $\Delta p = F\Delta t = m\Delta v$

           $\Delta p = (5.0)\,(10.0) = 5.0 \times 10^2$ N-s

     B. $\Delta p = m\Delta v$

           $5.0 \times 10^2 = (2.0)\Delta v$

           $\Delta v = 25$ m/s

Example 2   An automobile traveling at 100 km/h strikes a barrier
                 which brings the car to a halt in 1.5 s.
                 A. Find the impulse imparted on a 70 kg person in the car.
                 B. Find the force exerted on the person.
                 C. Find the force the person exerts on the seat belt.

Solution

     A. $\Delta p = m\Delta v = F\Delta t$

           $\Delta p = (70)(27.8)$ convert km/h to m/s

           $\Delta p = 1.9 \times 10^3$ N-s

     B. $\Delta p = F\Delta t$

           $1.9 \times 10^3 = F\,(1.5\text{ s})$

           1.3×10^3 N

     C.   The person would exert an equal an opposite force on the
           belt -Newton's third law of motion.

Conservation of Linear Momentum

Example 1   A 2.0 kg mass travels along the positive x-axis at 10.0 m/s. It collides with a 1.0 kg particle initially at rest. After the collision, the two stick together. Find their common velocity.

Solution

$m_1 = 2.0$ kg ; $m_2 = 1.0$ kg ; $v_{1o} = 10.0$ m/s + x

$p_{before} = p_{after}$

$m_1 v_{1o} + m_2 v_{2o} = (m_1 + m_2) v$

$(2.0)(10) + 0 = (3.0)v$

$v = 6.7$ m/s + **x**

Example 2   A 50.0 kg child initially at rest on a skateboard catches a 0.50 kg ball moving at 20.0 m/s. Find the speed of the child after he makes the catch.

Solution

$p_{before} = p_{after}$

$m_c v_{co} + m_b v_{bo} = (m_c + m_b) v$

$(50.0)(0) + (0.50)(20.0) = (50.5) v$

$10 = 50.5 v$

$v = 0.20$ m/s

Example 3   A 2.0 kg mass moving at 3.0 m/s in the positive x-direction collides elastically with a 5.0 kg mass initially at rest. Find the velocity of each mass immediately after the collision.

Solution

$m_1 = 2.0$ kg ; $m_2 = 5.0$ kg ; $v_{1o} = 3.0$ m/s

$v_2 = 2m_1 v_1 / (m_1 + m_2)$

$v_2 = 2 (2.0) (3.0) / (7.0)$

$v_2 = 1.7$ m/s +**x** direction

$$m_1 v_{1o} = m_1 v_1 + m_2 v_2$$

$$(2.0)(3.0) = (2.0)(v_1) + (5.0)(1.7)$$
$$6.0 = 2.0\, v_1 + 8.5$$
$$-2.5 = 2.0\, v_1$$
$$v_1 = -1.3 \text{ m/s or } 1.3 \text{ m/s in the } \textbf{-x} \text{ direction}$$

Example 4 A 4.0 kg mass moves along the **+y** axis at 2.0 m/s when it strikes a 5.0 kg mass moving at 1.0 m/s along the **-y** axis. After the collision the 5.0 kg mass moves at 0.5 m/s in the positive **+y** direction.

Solution

A. Find the velocity of the 4.0 kg mass after the collision.
B. Find the loss of mechanical energy.

let $m_1 = 4.0$ kg ; $m_2 = 5.0$ kg ; $v_{1o} = +2.0$ m/s ; $v_{2o} = -1.0$ m/s

$$p_o = p_f$$
$$m_1 v_{1o} + m_2 v_{2o} = m_1 v_1 + m_2 v_2$$
$$(4.0)(2.0) + (5.0)(-1.0) = (4.0)(v_1) + (5.0)(0.5)$$
$$8 - 5 \ = 2v_1 + 2.5$$
$$3 - 2.5 = 2v_1$$
$$v_1 = 0.25 \text{ m/s + y direction}$$

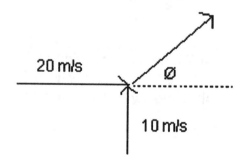

Example 5 A 1000 kg car approaches an intersection at 20 m/s East as an identical car travels 10 m/s North. At some point the cars collide and stick. Find their common velocity.

<u>Solution</u>

$m_1 =$ car which is traveling East $- v_{1o}$

$m_2 =$ car which is traveling North $- v_{2o}$

First work with momentum in the **x** component.

$$m_1 v_{1o} = (m_1 + m_2)\, v\, (\cos\emptyset)$$
$$(1000)(20) = (2000)\, v\, (\cos\emptyset)$$
$$10 = v \cos\emptyset$$

Then work with the momentum in the **y** component.

$$m_2 v_{2o} = (m_1 + m_2)\, v\, (\sin\emptyset)$$
$$(1000)(10) = (2000)\, v(\sin\emptyset)$$
$$5 = v\, (\sin\emptyset)$$

Now take the y equation and divide by the x equation.

$$5/10 = \text{Tan}\, \emptyset$$
$$\emptyset = 27° \text{ N of E}$$

Now substitute back into the **y** equation.

$$5 = v\, (\sin 27)$$
$$v = 11 \text{ m/s}$$

Solutions to Selected problems from the Text

4. given : $m = 0.010$ kg ; P (from problem 3 = 450 kg-m/s)

$$P = mv$$
$$v = P/m$$
$$v = (4.5 \times 10^2) / 1.0 \times 10^{-2}$$
$$v = 4.5 \times 10^4 \text{ m/s}$$

9. given : $m = 0.10$ kg ; $v_o = 0.50$ m/s ; $v_f = -0.50$ m/s

(opposite direction)

87

$$P = mv_f - mv_o$$
$$P = m\,(v_f - v_o)$$
$$P = (0.10)\,(-0.50 - 0.50) = -0.10 \text{ kg-m/s}$$

The significance of the negative sign shows the change is in the opposite direction of the initial velocity.

14. given :   w of the hunter = 750 N ; m_{canoe} = 30 kg ;
v_{hunter} = 1.5 m/s

first find the mass of the hunter w / g = m ;
$$m = 750 / 9.8 = 76.5 \text{ kg}$$

Momentum is conserved since the net external force is zero.

$$0 = m_c v_c + m_m v_m$$
$$0 = (30)(v_c) + (76.5)(1.5)$$
$$v_c = -3.8 \text{ m/s or 3.8 m/s away from the shore}$$

19. given : m_1 = mass of the block = 5.0 kg ;
m_2 = mass of bullet = 0.010 kg ;
v_{bo} = initial velocity of bullet = 300 m/s
v_b = final velocity of the bullet = 200 m/s

A.   Momentum is again conserved.

$$m_2 v_{bo} = m_2 v_b + m_1 v_1$$
$$(0.01)(300) = (0.01)(200) + 5\,(v_1)$$
$$3 - 2 = 5\,v_1$$
$$v_1 = 0.20 \text{ m/s in the initial direction of the bullet}$$

B. $W = \Delta K$
$$W = (1/2)m_1 v^2_1$$
$$W = (1/2)(5.0)(0.2)^2$$
$$W = 0.10 \text{ J}$$

C. $\Delta P = F \Delta t$

$\Delta t = \Delta P / F$

$\Delta t = (0.01)(100) / (0.10 / 0.10)$

$\Delta t = 1.0\ s$

24. given : m_1 = mass of one car = 1000 kg ; h = 4.5 m

First find the velocity of the car after it rolls down the hill and strikes the other car. Use the law of conservation of energy.

PE = K

$mgh = (1/2)mv_o^2$

$(9.8)(4.5) = (1/2)\ v_o^2$

$v_o = 9.4\ m/s$

When the cars collide momentum is conserved.

$m_1 v_{o1} = 2m_1\ v$

since the mass doubles the velocity is cut in half.

$v = 4.7\ m/s$ in the direction of m_1

30. given :   masses of the cars are identical.

A. v_1 = slow car = 10 km/h ; v_2 = fast car = 20 km/h

$mv_1 + mv_2 = 2\,m\,v$

$(m)(10) + m(20) = 2\,m\,(v)$

$v = 15\ km/h = 4.2\ m/s$

B. $K_{before} = (1/2)m(2.8)^2 + (1/2)m(5.6)^2$

$K_{before} = 19.6\ (m)$

$K_{after} = (1/2)(2m)(4.2)^2$

$K_{after} = 17.6\ (m)$

Therefore K is not conserved !

Review Questions

1. Why is impulse equal to the change in momentum ?

   By definition, impulse = $F \Delta t$, and using the relationships,
   $F = ma = \Delta v / \Delta t = (v_f - v_o) / \Delta t$, we have impulse = $p_f - p_o = \Delta p$.

2. Is the unit of momentum N-s or kg-m/s ?

   Both since the units are equivalent. N-s = $(kg$-$m/s^2)$ s = kg-m /s/

3. When jumping from a height, why does a person bend the knees on landing ?

   To increase the contact time and reduce the force to try avoiding injury. The impulse is a constant or equal to the person's momentum just before striking the ground and coming to a halt.

4. If there is an external force, why isn't momentum conserved ?

   Force is equal to the time rate of change of momentum, and if the momentum is changing it is not conserved.

5. If the total momentum of a system is conserved, are the individual momenta of the particles of the system conserved ?

   Not necessarily. The individual momenta may change as long as they vectorially add up to the total constant momentum at any instant.

6. What is the general definition of a collision ?

   A collision is any interaction in which momentum is exchanged or transferred. Since momentum is exchanged, energy is also exchanged - $K = (1/2)mv^2 = p^2 / 2m$.

7. What two totals are conserved in an elastic collision ?

The total linear momentum and the total energy. The total
energy of a system is conserved in every instance.

8. If two particles of mass *m* collided head-on with equal speeds, what
would be the situation after the collision if the collision were elastic ?
inelastic ?

The particles would return along their original paths such that the
momentum is zero (equal and opposite as before the collision).
If the collision were elastic, the particles would have the same
speed as before the collision. In an inelastic collision, the
particles would have the same speeds, but different from
the initial speeds since some energy would have been lost.

9. Why are internal forces not a consideration in the conservation of
momentum ?

They are, however, being internal forces and occurring in equal
and opposite pairs (Newton's third law of motion) they cancel
out. They are commonly ignored.

10. What is reverse thrust ?

A procedure by which a rocket or engine exhaust is directed
oppositely so the thrust is reversed, and is commonly used for
braking action for jet planes and spacecraft.

Sample Quiz

(Remove the quiz from the book and test your knowledge of the chapter material as though you were taking an in-class quiz. Check your answers with the key at the back of the Study Guide.)

Completion

1. The automobile air bag reduces injury through increasing the_____.

2. A ball of mass m and traveling with a horizontal speed v is caught by a player. The impulse imparted to the player's glove or hand is equal to _____.

3. If a particle is accelerating, then its linear momentum is_____.

Multiple Choice.

___ 4. The total momentum of a system is conserved
A. if there are no internal forces
B. there is an external force
C. only if the kinetic energy is conserved
D. none of the above

___ 5. When reverse thrust is used to brake a jet plane, the exhaust gases are directed
A. rearward B. upward   C. forward   D. downward

Problems

6. A 0.010 kg particle traveling with a speed of 4.0 m/s has a collision and leaves with a speed of 3.0 m/s in the opposite direction. What is the impulse ?

7. A ball of mass 0.25 kg and traveling with a speed of 10 m/s collides elastically head-on with a stationary ball of mass 0.75 kg.
    A. What are the momenta of the balls after the collision ?
    B. What is the speed of each mass after the collision ?
    C. What is the total momentum ?

8. A 1.0 kg mass traveling at 5.0 collides with a stationary 2.0 kg mass. After colliding the masses stick together, what is their common speed ?

Chapter 10

Rotational Motion and Dynamics

Sample Problems

Description of Rotational Motion

Example 1   A merry-go-round has a radius of 3.0 m and an angular
speed of 20 rpm.
A. Find the angular speed of the merry-go-round in rad/s.
B. Find the tangential speed of the outer edge of the
merry-go-round.

Solution

A. ω = (20 rev/min) (2 π rad / 1 rev) (1 min / 60 s) = 2.1 rad/s

B. $v = \omega r$
$v = (2.1)(3.0)$ (angular velocity must be
measured in rad/s)
$v = 6.2$ m/s

Example 2   A cylinder, initially at rest, undergoes an angular
acceleration of 2.0 rad/s^2 for 2.0 s.
A. Find the angular velocity of the cylinder after 2.0 s.
B. Find the angular displacement of the cylinder in radians
and in revolutions during the 2.0 s interval.

Solution

A.   given : $\omega_o = 0$; $\alpha = 2.0$ rad/s^2 ; t = 2.0 s

$\omega_f = \omega_o + \alpha t$

$\omega_f = 0 + (2.0)(2.0) = 4.0$ rad/s

B. $\theta = \omega_0 t + (1/2)\alpha t^2$

$\theta = (0)(2.0) + (1/2)(2.0)(2.0)^2$

$\theta = 4.0$ rad

Example 3   A car starts from rest and accelerates at the rate of 2.0 m/s^2 for 10 s. If the radius of the tires is 0.30 m and the tires do not spin or slip, find
A. the angular velocity of the tires after 10.0 s.
B. the number of revolutions the tires make during the 10.0 s interval.
C. the angular acceleration of the wheel.
D. the total acceleration of a point on the edge of the wheel after 10.0 s.

Solution

Treat the linear quantities separate from the angular quantities.

given : $a_t = 2.0$ m/s^2 ; $v_o = 0$; t = 10.0 s

$\omega_o = 0$; t = 10.0 s

A. find the linear velocity after 10.0 s.

$v = v_o + at$

$v = 0 + (2.0)(10.0) = 20.0$ m/s

next find the angular speed

$v = \omega r$

$(20.0) = \omega (0.30)$

$\omega = 67$ rad/s

B. $a_t = \alpha r$

$\alpha = (2.0) / 0.30$

$\alpha = 6.7$ rad/s^2

95

$$\theta = \omega_0 t + (1/2)\alpha t^2$$
$$\theta = 0 \quad + (1/2)(6.7)(10.0)^2$$
$$\theta = (3.3 \times 10^2 \text{ rad}) \ (1 \text{ rev} / 2\pi \text{ rad}) = 53 \text{ rev}$$

C.   found in part B

D. $a_t = 2.0 \text{ m/s}^2$

$$a_c = v^2/r$$
$$a_c = (20)^2 / 0.3 = 1.3 \times 10^3 \text{ m/s}^2$$
$$a^2_{total} = a_c^2 + a_t^2$$
$$a = 1.3 \times 10^3 \text{ m/s}^2 \quad \text{(2.0 is insignificant compared to } a_t \text{)}$$

Torques and Moment of Inertia

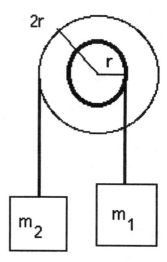

Example 1   In the figure above, $m_1 = 2.0$ kg. Find the mass m_2 needed to keep the system at rest.

Solution

$$\tau_{clockwise} = \tau_{counterclockwise}$$
$$F_1 r_1 = F_2 r_2$$

$$(m_1g)(r_1) = (m_2g)(r_2)$$
$$(2.0)(r) = (m_2)(2r)$$
$$r = 1.0 \text{ kg}$$

Example 2   The system in the previous example remains in equilibrium when m1 is 2.0 kg, $m_2 = 1.2$ kg, $r_1 = 10$ cm and $r_2 = 20$ cm. The system is at rest because of a frictional torque. Calculate the magnitude of the frictional torque.

Solution

$$(m_1g) \, r_1 + \tau_f = (m_2g) \, (r_2)$$
$$(2.0)(9.8)(0.10) + \tau_f = (1.2)(9.8)(0.20)$$
$$1.96 + \tau_f = 2.35$$
$$\tau_f = 0.39 \text{ m-N}$$

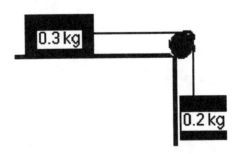

Example 3   The 0.30 kg mass is free to move on the frictionless surface. The pulley has a mass of 0.20 kg and a radius of 20.0 cm. The system is released from rest.   The moment of inertia for the pulley is $(1/2)mr^2$
A. Find the linear acceleration of the system.
B. Find the angular acceleration of the pulley.
C. Find the distance the 0.2 kg mass will descend in 2.0 s.

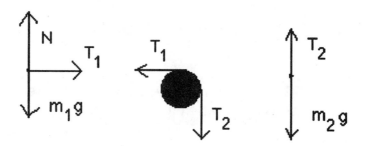

A. First apply Newton's second law of motion to the two bodies.

$m_1 a = T_1$

$m_2 a = m_2 g - T_2$

Now work with torques.

$\tau_{net} = T_2 r - T_1 r$

$I\alpha = T_2 r - T_1 r$

$(1/2) m_p r^2 \alpha = T_2 r - T_1 r$

$(1/2) m_p r \alpha = T_2 - T_1$

$(1/2) m_p a = T_2 - T_1$

$(1/2) m_p a + m_1 a + m_2 a = m_2 g$

$(1/2)(0.2) a + (0.3) a + (0.2) a = (0.2)(9.8)$

$0.6 a = 1.96$

$a = 3.3 \, m/s^2$

B. $a_t = \alpha r$

$\alpha = a_t / r$

$\alpha = (3.3) / (0.20)$

$\alpha = 17 \, rad/s^2$

C. $x = v_0 t + (1/2) a t^2$

$x = 0 \; + (1/2)(3.3)(2.0)^2$

$x = 6.6 \, m$

Rotational Work, Power, and Kinetic Energy

Example 1　A net torque of 10.0 m-N acts on a cylinder, mass 2.0 kg, radius 0.30 m for 10.0 s. At the end of the 10.0 s interval,
　　　A. find the angular velocity.
　　　B. find the kinetic energy of the cylinder.
　　　C. calculate the net work.
　　　D. calculate the angular displacement.

Solution

A.　$\tau = 10.0$ m-N ; $m = 2.0$ kg ; $r = 0.30$ m ; $t = 10.0$ s
　　$\tau = I\alpha$; $I = (1/2)mr^2 = (1/2)(2.0)(0.30)^2 = 0.090$ kg-m^2
　　$10.0 = (0.090)\alpha$
　　$\alpha = 1.1 \times 10^2$ rad/s^2

B.　$\omega_f = \omega_0 + \alpha t$
　　$\omega_f = 0 + (1.1 \times 10^2)(10.0) = 1.1 \times 10^3$ rad/s
　　$K = (1/2)\, I\omega^2 = (1/2)(0.09)(1.1 \times 10^3)^2$
　　$K = 5.6 \times 10^4$ J

C.　$W = \Delta K = 5.4 \times 10^4$ J

D.　$W = \tau\, \theta$
　　$\theta = (5.4 \times 10^4) / 10.0 = 5.4 \times 10^3$ rad

Example 2　A sphere is rolling at 5.0 m/s before it moves up an inclined plane. How high will the sphere move up the plane ?

Solution
　　$v = 5.0$ m/s

　　Energy is conserved. The total kinetic energy at the bottom (sum of the rotational and translational) equals the potential energy at the top.

$$KE_T + KE_R = PE$$

$$(1/2)mv^2 + (1/2)\, I\omega^2 = mgh$$

$$(1/2)mv^2 + (1/2)(2/5\; mr^2)(v/r)^2 = mgh$$

$$(1/2)mv^2 + (1/5)mv^2 = mgh$$

$$(7/10)mv^2 = mgh$$

$$(7/10)(5.0)^2 = (9.8)\, h$$

$$h = 1.8\ m$$

Angular Momentum

Example 1   A net torque of 20.0 m-N is applied to a cylinder whose mass is 1.0 kg and radius is 40.0 cm for 30.0 s.
A. Calculate the change in angular momentum.
B. Calculate the change in angular velocity.

Solution

$\tau = 20.0$ m-N ; $I = (1/2)\, mr^2$; $r = 0.40$ m; m=1.0 kg;t = 30.0 s

A. $\tau = \Delta L / t$
$20.0 = \Delta L / 30.0$
$\Delta L = 6.0 \times 10^2$ m-N

B. $\Delta L = I\, \Delta\omega$
$6.0 \times 10^2 = (1/2)(1.0)(0.40)^2 \Delta\omega$
$\Delta\omega = (6.0 \times 10^2) / 0.08 = 7.5 \times 10^3$ rad /s

Example 2   A diver jumps off the high board. He has a moment of inertia of 60 kg-m^2 when in a tuck position and 90 kg-m^2 when in a straight position. When the diver is in a straight position he has an angular speed of 2 rps. Find his angular velocity when he is in the tuck position.

100

<u>Solution</u>

Since the net torque is zero angular momentum is
conserved.

$$L_o = L$$

$$I_1 \omega_1 = I_2 \omega_2$$

$$(60) \, \omega_1 = (90) \, (2 \text{ rps})$$

$$\omega_1 = 3.0 \text{ rps}$$

<u>Solutions to Selected problems from the Text</u>

6. given :   angular displacement $= \pi$ radians ; $t = 30.0$ s

$$\omega = \theta / t$$

$$\omega = \pi / 30 = (3.14) / 30$$

$$\omega = 0.105 \text{ rad/s}$$

11. given :   initial angular velocity $= 1.6$ rad/s ; time $= 5.0$ s ; angular

acc.$= 0.80$ rad/s^2

$$\theta = \omega_o t + (1/2) \alpha t^2$$

$$\theta = (1.6)(5.0) + (1/2)(0.80)(5.0)^2$$

$$\theta = 18 \text{ rad} \ (1 \text{ rev} / 2 \, \pi \text{ rad}) = 2.9 \text{ rev}$$

16. given :   torque $= 20$ lb-ft ; force $= 15$ lb ; angle between torque and

force $= 37°$

A. $r_{perpendicular} = \tau / F$

$r_{perpendicular} = 20 / 15 = 1.3$ ft

B. $\tau = r \, F \sin \theta$

$20 = (r) \, (15) \, (\sin 37)$

$20 = 9 \, r$

$r = 2.2$ ft

101

20. given : initial angular velocity = 0 ; final angular velocity =1500 rpm ;
     time = 5.0 s ; moment of inertia = 40 kg-m^2

   First find the angular acceleration.

   $\omega_f = \omega_o + \alpha t$

   $157 = 0 + (\alpha)(5.0)$

   $\alpha = 31.4$ rad/s^2

   $\tau = I\alpha$

   $\tau = (40)(31.4) = 1.3 \times 10^3$ m-N

25. given :   radius of the sphere = 0.10 m

   A.   First use Newton's second law of motion.

      $ma = mg - 2T$

      The use torques.

      $\tau = I\alpha$; $\tau = 2T r$

      $I\alpha = 2T r$

      $(1/2)mr^2\alpha = 2T r$

      $(1/2)ma = 2 T$ since a = αr
      combine with the equation for Newton's second law
      $(1/2)ma + ma = mg - 2T + 2T$
      $1.5 a = g$ (assume m are equal)
      $a = 6.5$ m/s^2
      $\alpha = 6.5$ m/s^2 / 0.10 m = 65 rad /s^2

   B. $\omega_f^2 = \omega_o^2 + 2\alpha\theta$
      $\omega_f^2 = 0^2 + 2(65)(25.1)$
      $\omega = 57$ rad /s

30. given :   mass = 6.0 kg ; radius 0.80 m ; angular speed = 20 rad/s

$$K = (1/2)\ I\ \omega^2$$
$$K = (1/2)\ (MR^2)\ \omega^2$$
$$K = (1/2)(6.0\ )(0.80)^2\ (20)^2$$
$$K = 7.7 \times 10^2\ J$$

34. given :   height = 4.0 m
The potential energy at the top of the inclined plane equals the total kinetic energy (translational and rotational) at the bottom of the inclined plane.

$$PE = K_R + K_T$$
$$mgh = (1/2)(MR^2)\ \omega^2 + (1/2)\ mv^2$$
$$mgh = (1/2)\ mv^2 + (1/2)mv^2 \qquad\qquad v = \omega r$$
$$mgh = mv^2$$
$$(9.8)(4.0) = v^2$$
$$v = 6.3\ m/s$$

39. given   : R is the radius of the loop

A.   At the top , the sphere is moving in a circle. The weight supplies the centripetal force. Any speed greater than this will require the track to push downward.

$$mv^2/R = mg$$
$$v = (rg)^{1/2}$$

B.   Using the law of conservation of energy :
$$P_{beginning} = P_{top} + K_{translational} + K_{rotational}$$
$$mgh = mg(2r) + (1/2)mv^2 + (1/2)(2/5mr^2)\omega^2 r^2$$
$$mgh = 2\ mgr + 7/10\ mv^2$$
$$mgh = 2\ mgr + (7/10)\ m\ rg$$
$$mgh = 2.7\ mgr$$
$$h = 2.7\ R$$

103

45. given : initial angular momentum = 80 m-N-s ; final angular
momentum
100 m-N-s   ; t = 5.0 s

$\tau = \Delta L / t$

$\tau = (100 - 80) / 5.0$

$\tau = 4.0$ m-N

Review Questions

1. What is meant by angular distance ?

This is the angle through which something rotates or revolves in
rotational motion. It is expressed in degrees or in radians and is
related to the arc length by $s = r\theta$ where θ is in radians.

2. How is the angular velocity related to the tangential velocity ?

By the relationship $v = r\omega$. Note that the tangential speed or
velocity of a particle or a rotating rigid body varies with r. The
angular velocity is the same for all particles of a rotating body.

3. Distinguish between angular acceleration and tangential
acceleration.

Angular acceleration is the time rate of change of angular
velocity and is related to the tangential acceleration by $a_t = r\alpha$. If
there is an angular acceleration, then there is a tangential
acceleration. This increases the tangential velocity and requires
an increase in the centripetal acceleration of a particle to remain
in the circular path ($a_c = v^2/r$). The net acceleration is the vector
sum of the tangential and centripetal accelerations.

4. What is the moment of inertia for a particle in circular motion ?

$I = mr^2$, where r is the radius of the circular path.

5. Give the rotational analogs for (a) mass, (b) Newton's second law of motion, and (c) work.

   (a) moment of inertia - I
   (b) $\tau = I\alpha$
   (c) $W = \tau\theta$

6. How may the moment of inertia of an object be increased ?

   By increasing the mass and / or shifting the mass distribution farther away from the axis of rotation.

7. Explain why $s = r\theta$ is a condition for rolling without slipping.

   If an object rolls without slipping, then the linear distance s it rolls is the same as the circumference arc length through which it rolls. If slipping occurs, then s is greater than the linear distance d. Could d ever be greater than s ?

8. What kinds of energy does a rolling object have ?

   If rolling on a level surface, the object would have both translational and rotational kinetic energy. If on an inclined plane, it would also have potential energy.

9. How is angular momentum related to torque ?

   Torque is the time rate of change of angular momentum, $\tau = \Delta L / \Delta t$.

10. When is angular momentum conserved, and if so, is the angular velocity always constant ?

    The angular momentum is conserved in the absence of a net external torque. Since $L = I\omega$, L is constant as long as the product of I and ω are constant, so ω can be changes by changing I.

Sample Quiz

(Remove the quiz from the book and test your knowledge of the chapter material as though you were taking an in-class quiz. Check your answers with the key at the back of the Study Guide.)

1. For a rotating body, the instantaneous angular_____, _____, and _____ are the same for each particle of the body.

2. For a rolling body, if $v \neq r\omega$, the object _____.

3. The stability of a two-wheeled bicycle is based on the_____ _____ .

Multiple Choice.

___ 4. Technically, the unit of angular acceleration is
   A. m/s^2 B. $kg\text{-}m/s^2$ C. rad/s^2 D. s^{-2}

___ 5. If a torque acts on a rotating body, then there is no change in the
   A. angular acceleration   C. kinetic energy
   B. moment of inertia   D. angular momentum

Problems

6. The angular speed of a rotating disk increases uniformly from 2.5 rad/s to 5.5 rad/s while the disk makes 2 revolutions. If the disk has a moment of inertia of 3.0 kg-m^2, what is the magnitude of the torque that acts on the disk ?

7. A hoop ($I = mr^2$) rolls down an inclined plane with a vertical height of 2.0 m without slipping. What is the linear speed of the hoop when it rolls onto the horizontal surface ?

Chapter 11   Machines

Sample Problems

Mechanical Advantage and Efficiency

Example 1   A machine is designed for an input force of 100 N and an output force of 150 N. The input distance is 5.0 cm and the output distance is 2.5 cm
A. Find the IMA.
B. Find the AMA.
C. Calculate the efficiency of the machine.

Solution

$$F_i = 100 \text{ N} \; ; \; F_O = 150 \text{ N} \; ; \; d_i = 5.0 \text{ cm} \; ; \; d_O = 2.5 \text{ cm}$$

A. $IMA = d_i / d_O$
      $IMA = 5.0 / 2.5 = 2.0$

B. $AMA = F_O / F_i$
      $AMA = 150 / 100 = 1.5$

C. $Eff = AMA / IMA$
      $Eff = 1.5 / 2.0 = 0.75$ or 75 %

Example 2   The efficiency for a machine is 50%. If the machine has an IMA of 5.0 and can lift a maximum of 1500 N, find the
A. AMA
B. the input force.

Solution

A. $Eff. = AMA / IMA$
      $0.50 = AMA / 5.0$
      $AMA = 2.5$

B. $AMA = F_o / F_i$

  $2.5 = 1500 \text{ N} / F_i$

  $F_i = 600 \text{ N}$

Simple Machines

Example 1   An inclined plane 3.0 m long is 0.5 m high. A worker must exert a force of 100 N to move a 400 N weight up the plane.
A. Calculate the IMA.
B. Calculate the AMA.
C. Calculate the efficiency.

Solution

A. $IMA = L / h$

  $IMA = 3.0 / 0.5$

  $IMA = 6.0$

B. $AMA = F_o / F_i$

  $AMA = 400 / 100 = 4.0$

C. $Eff = AMA / IMA$

  $Eff = 6.0 / 4.0$

  $Eff = 1.5$

Example 2   A force of 600 N is needed to move an object weighing 800 N using a pulley system. When the rope is pulled 1.0 m, it causes the weight to move 10 cm.
A, Calculate the IMA.
B. Calculate the AMA.
C. Calculate the efficiency.

Solution

$F_o = 800 \text{ N} ; F_i = 600 \text{ N} ; d_i = 1.0 \text{ m} ; d_o = 0.10 \text{ m}$

A. $IMA = d_o / d_i$

  $IMA = 1.0 / 0.10 = 10$

B. $AMA = F_o / F_i$
    $AMA = 800 / 600 = 1.3$

C. $Eff = AMA / IMA$
    $Eff = 1.3 / 10 = 0.13 = 13\%$

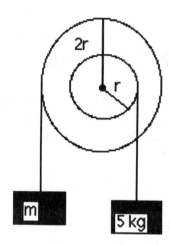

Example 3 In the figure above, find the
           A. mass m needed to lift the 5.0 kg mass.
           B. IMA for the wheel and axle.

Solution

    A. $m_1 gr = m_2 g\,(2r)$
        $(5.0)\,r = m\,(2r)$
        $m = 2.5\,kg$

    B. $IMA = R/r$
        $IMA = 2r / r$
        $IMA = 2$

Example 4 A screw jack has a lever arm of 0.50 m. There are 2.0
           threads per 1.5 cm.   A force of 200 N is needed to raise a
           load of 400 N.
           A. Calculate the IMA.
           B. Calculate the AMA.
           C. Calculate the efficiency.

110

<u>Solution</u>

A. $IMA = d_i / d_o$

$IMA = 2\pi r / \rho$ $\rho = 1 / (2.0 / 1.5) = 0.75$
$IMA = 2\pi (0.5) / 0.75$
$IMA = 4.2$

B. $AMA = F_o / F_i$
$AMA = 400 / 200 = 2$

C. $Eff = AMA / IMA$
$Eff = 2 / 4.2 = 0.48$ or 48%

Solutions to Selected problems from the Text

6. given : $d_{output} = 4.0$ cm ; $d_{input} = 12$ cm ; $F_{output} = 270$ N

$IMA = F_o / F_i$
$IMA = d_i / d_o$
$F_i = (F_o) (d_o) / d_i$
$F_i = (270) (4.0) / 12$
$F_i = 90$ N

7. given : $AMA = 2.4$
$IMA = d_i / d_o = 12 / 4.0 = 3.0$
$Eff = AMA / IMA$
$Eff = 2.4 / 3.0$
$Eff = 0.80$ or 80%

12. given : $L_{input} = 6.0$ cm ; $L_{output} = 36$ cm ; $m = 2.0$ kg
  A. $IMA = d_i / d_o = 6.0 / 36 = 0.17$
             $IMA = F_o / F_i$
             $F_i = (2.0)(9.8) / 0.17$
             $F_i = 118$ N
  B.   third class

111

17. given : $F_{output} = 22 \, lb$; $F_{input} = 20 \, lb$;

   A. $AMA = F_i / F_o$
            $AMA = 20 / 22$
            $AMA = 0.91$
   B. $Eff = AMA / IMA$
            $Eff = 0.91 / 1.0$
            $Eff = 0.91$ or 91 %

22. given : $F_{input} = 200 \, lb$; $F_{output} = 800 \, lb$
   A. $AMA = F_o / F_i$
            $AMA = 800 / 200$
            $AMA = 4.0$

   B. $Eff = AMA / IMA$
            $Eff = AMA \, (\sin 10)$
            $Eff = 0.69$

27. given : length = 4.0 cm ; R = 2.5 cm ; r = 0.5 cm

            $IMA = L / (R - r)$
            $IMA = 4.0 \, / \, (2.5 - 0.5)$
            $IMA = 4.0 / 2.0$
            $IMA = 2.0$

33. given : $a_{output} = 20 \, in^2$; $a_{input} = 2.0 \, in^2$; $L_i = 8.0 \, in$; $L_o = 2.0 \, in$

   A. $IMA_1 = A_o / A_i$
            $IMA_1 = 20 / 2.0 = 10$
            $IMA_2 = L_i / L_o$
            $IMA_2 = 8.0 / 2.0 = 4.0$
            Total IMA = 40

   B. $Eff = AMA / IMA$
            $Eff = (F_o / F_i) / IMA$
            $Eff = (500 / 15) / 40$
            $Eff = 0.83$ or 83%

112

39. given : radii 6.0 in , 4.0 in , and 2.0 in ; $\omega_0 = 1200$ rpm

$$\omega_0 / \omega_i = D_i / D_o$$

Reductions (2/6) : 400 rpm

(2/4) : 600 rpm

(4/6) : 800 rpm

Increases (6/4) : 1800 rpm

(4/2) : 2400 rpm

(6/2) : 3600 rpm

44.

$$SR_1 = N_3 / N_2$$
$$SR_1 = 46 / 18$$
$$SR_2 = N_2 / N_1$$
$$SR_2 = 18 / 12$$
$$\omega_0 = [(SR_1) / (SR_2)] \omega_1$$
$$\omega_0 = (46 / 12) (15 = 58 \text{ rpm}$$

Review Questions

1. Do machines multiply or increase work ?

   No, machines multiply force. The work output is always less than the work input due to frictional losses.

2. When a machine multiplies force, what else is affected ?

   Force multiplication is done at the expense of a distance reduction.

3. What is the difference between IMA and AMA ?

The ideal mechanical advantage (IMA) neglects frictional and other losses, and the actual mechanical advantage (AMA) takes losses into account.

4. What is an advantage in finding the IMA of a machine rather than the AMA ?

The IMA can be computed from geometrical considerations, i.e., determined theoretically. The AMA must be determined from experimental considerations.

5. What are the simple machines that use (a) the level principle and (b) the inclined plane principle ?

(a) the lever, the pulley, and the wheel and axle
(b) the inclined plane, the wedge, and the screw

6. Which class of lever is not a force multiplier ?

Third class in which input force is between the fulcrum and the output force (shovel). The load is moved through a greater distance at the expense of a force reduction.

7. What is the principle of the hydraulic press ?

Pascal's principle and pressure (F / A). By varying the input and output piston areas, the input and out-up forces may be varies, in particular, the output force multiplied.

8. Why is a movable pulley a "lever" and a force multiplier ?

A movable pulley is essentially a second-class level (output force between the fulcrum and input force) , and with unequal lever arms can multiply force. A fixed pulley on the other hand, has equal lever arms and does not multiply force, but acts as a direction changer.

9. When a machine is used to transmit power, what does it multiply in this case ?

In general, torque is multiplied. Also, the angular speed may be multiplied at the expense of a torque reduction.

10. What is the speed ratio and how is it related to pulley drives and gear drives ?

The speed ratio (SR) is the ratio of the output and input angular speed (ω_o / ω_i), and hence is the "speed multiplier" factor. For a pulley drive, the SR is the gear ratio of the input and output pulley diameters (D_i / D_o), and for a gear drive, the ratio of the input and output number of gear teeth (N_i / N_o).

Sample Quiz

(Remove the quiz from the book and test your knowledge of the chapter material as though you were taking an in-class quiz. Check your answers with the key at the back of the Study Guide.)

Completion

1. The AMA is the ratio of F_o / F_i and the IMA is the ratio of _____.

2. In terms of AMA and TMA, the efficiency of a machine is equal is _____.

3. In one revolution, a screw moves a lateral distance equal to the _____ of the screw.

Multiple Choice

_____ 4. The IMA of an inclined plane may be increased by
   A. making the plane more smooth
   B. increasing its height
   C. increasing the AMA
   D. decreasing the angle of incline

_____ 5. If the number of teeth on an input gear is less than the number of teeth on an output gear for a gear drive, then
   A. the output speed is less than the input speed
   B. the speed ratio is greater than one
   C. the IMA is equal to one
   D. none of the above

Problems

6. If the input and output forces for moving an object up a 30° inclined plane are 200 lb and 350 lb, respectively, what is the efficiency of the plane ?

7. A pulley system output pulley with a diameter of 4.0 cm and a drive input pulley with two step pulleys with diameters of 2.0 cm and 5.0 cm. If the input speed is 150 rpm, what are the possible output speeds ?

Chapter 12   Mechanical Properties of Materials

Sample Problems

Stress and Strain

Example 1   A 100 kg mass is hung from a copper wire whose length is 2.0 m and radius is 2.0 mm. Find the stress in the wire.

Solution

$m = 100$ kg therefore the weight is 980 N ; $r = 2.0$ m

Area $= \pi r^2$
$A = \pi (2.0 \times 10^{-3} \text{ m})^2 = 1.3 \times 10^{-5} \text{ m}^2$

Stress $= F / A$
Stress $= 9.8 \times 10^2 / 1.3 \times 10^{-5}$
Stress $= 7.5 \times 10^7 \text{ N/m}^2$

Example 2   Find the greatest load a wire in the previous example can withstand without breaking ?

Solution

Tensile strength $= 3.4 \times 10^8 \text{ N/m}^2$

$TS = F / A$
$(TS)(A) = F$
$(3.4 \times 10^8)(1.26 \times 10^{-5} \text{ m}^2) = F_{max}$
$4.4 \times 10^3 \text{ N} = F_{max}$
$4.5 \times 10^2 \text{ kg} = m_{max}$

Young's Modulus

Example 1   A mass of 100 kg is placed at the end of a rod whose
              length is 50 cm and diameter 2.0 cm.
              A. Find the stress in the rod.
              B. Find the strain in the rod.

Solution
      A.   Stress = F / A
              Stress = (100)(9.8) / π (1.0 x 10^{-2})^2
              Stress = 3.1 x 10^6 N/m^2

      B.   Y = stress / strain
              20 x 10^{10} = 3.1 x 10^6 / strain
              strain = 1.55 x 10^{-5}

Example 2   A bone 10 cm long has a cross-sectional area of 2.8 cm^2.
              The ultimate compressive force for the bone is
              1.6 x 10^8 N/m^2. Find the compressive force needed to
              break the bone.
Solution

              F_{max} = 1.6 x 10^8 N/m^2 ; A = 2.8 cm^2 = 2.8 x 10^{-4} m^2
              F_{max} = (1.6 x 10^8)(2.8 x 10^{-4}) = 4.5 x 10^4 N

Shear Modulus

Example   A torques of 50 m-N is applied to an aluminum cylinder
              whose diameter is 4.0 cm and length 10.0 cm.   Find the
              twist angle.

Solution

              τ = 50 m-N ; r = 2.0 cm ; length = 10.0 cm

              S = 2τL / π θ r^4

$$\theta = 2\tau L / \pi r^4 S$$
$$\theta = 2\,(50)\,(0.10) / \pi(0.02)^4(2.4 \times 10^{10})$$
$$\theta = 8.3 \times 10^{-4}\ rad$$

Bulk Modulus

Example   A pressure of 30 lb/in^2 is applied to an iron cube whose sides measure 4.0 in.   Find the change in volume of the cube.

Solution

$$V = (4.0\ in)^3 = 64\ in^3$$

$$B = p / (\Delta V / V_o)$$
$$\Delta V = VP / B$$
$$\Delta V = (64\ in^3)(\ 30\ lb/in^2) / (1.45 \times 10^6\ lb/in^2)$$
$$\Delta V = 1.32 \times 10^{-3}\ in^3$$

Solutions to Selected problems from the Text

5. given : F = 80 N ; dimensions of the block 0.30 m x 0.50 m

First calculate the area of the block : A = l w
$$A = (0.30)\,(0.50) = 0.15\ m^2$$
Stress = F / A
Stress = (80) / 0.15
Stress = 5.3×10^2 N/m^2

10. given : F = 250 lb ; elastic limit (EL) = 1.9×10^4 lb/in^2

EF = F / A
A = F / EF
$$A = (250) / 1.9 \times 10^4$$
$$A = 1.3 \times 10^{-2}\ in^2$$

$$A = \pi d^2 / 4$$
$$1.3 \times 10^{-2} = \pi d^2 / 4$$
$$5.2 \times 10^{-2} = \pi d^2$$
$$d = 0.13 \text{ in}$$

15. given : $Y_{steel} = 29 \times 10^6 \text{ lb/in}^2$; $L_0 = 15$ ft ; $a = 10 \text{ in}^2$;

$$F = 5.0 \times 10^4 \text{ lb}$$

$$Y = (F_n / A) / \Delta L / L_0$$
$$29 \times 10^6 = (5.4 \times 10^4 / 10) / \Delta L / 15$$
$$\Delta L = 2.6 \times 10^{-3} \text{ ft} = 0.031 \text{ in}$$

20. given : mass = 50 kg ; r = 1.0 cm ;

First work with forces : $F_v = (50 \text{ kg}) (9.8) = 490 \text{ N}$

Cables at angles have same force. $F_v = 2 \, T \sin 15$
$$490 = 2 \, T \, (\sin 15)$$
$$245 / \sin 14 = T$$
$$T = 1.0 \times 10^3 \text{ N}$$

$$Y = (F_n / A) / \Delta L / L_0$$
$$20 \times 10^{10} = (490 / 3.14 \times 10^{-4}) / (\Delta L / L)$$
$$\Delta L / L = 7.8 \times 10^{-6} \qquad \text{for the vertical cable}$$

$$20 \times 10^{10} = (1.0 \times 10^3 / 3.14 \times 10^{-4}) / (\Delta L / L_0)$$
$$\Delta L / L_0 = 1.6 \times 10^{-5}$$

25. given : torque = 100 in-lb ; L = 2.5 in ; diameter = 0.50 in

$$S = 2\tau L / \theta \, \pi r^4$$
$$5.1 \times 10^6 = 2 (100) (2.5) / \theta (\pi) (0.25)^2$$
$$5.1 \times 10^6 = 4.1 \times 10^4 / \theta$$
$$\theta = 4.1 \times 10^4 / 5.1 \times 10^6 = 8.0 \times 10^{-3} \text{ rad}$$

29. The shear angle is the angle Ø measured along the circumference of the free end and s = LØ, where L is the length of the rod. Since the twist Ø (at the end of the rod) is s = r θ, then Ø = (r / L) θ

36. given :   force $= 2.5 \times 10^7$ N ; A $= 0.06^2 = 3.6 \times 10^{-3}$ m^2

$$Y = (F_n / A) / \Delta L / L_o$$

$6.1 \times 10^{10} = (2.5 \times 10^7 / 3.6 \times 10^{-3}) / \Delta L / L$

$6.1 \times 10^{10} = 6.9 \times 10^9 / (\Delta L / L)$

$\Delta L / L = 6.9 \times 10^9 / 6.1 \times 10^{10}$

$\Delta L / L = 0.11$

Review Questions

1. Distinguish between stress and strain.

> Stress comes first, then strain. Stress is the applied force per area acting on a body.   Strain is the relative change in dimensions of the body resulting from a stress.

2. Is elastic deformation the same as plastic deformation ?

> NO. Elastic deformation does not exceed the elastic limit and a material recovers after the stress is removed, Plastic deformation occurs after the elastic limits exceeded and the deformation is permanent. Metal stamping is a good example.

3. What is the difference between yield strength and tensile strength ?

> Yield strength is the stress value at which appreciable plastic deformation begins (just beyond the elastic limit). Tensile strength is the maximum stress a material can support just before fracturing.

4. In general, what is the elastic modulus ?

The stress divided by the strain.

5. What is meant by a material having anisotropic moduli ?

The moduli are not the same in all directions. For example, wood has different longitudinal moduli along and across the wood grain.

6. Do elastic moduli hold for any amount of applied stress ?

NO. As the name implies, they are applicable for only elastic deformations, i.e. the stress is directly proportional to the strain (below elastic limit).

7. What is work or strain hardening ?

Some materials become harder and more brittle as a result of repeated flexing or shear stressing. As a result the ultimate shear strength is reduced. A common example is the breaking of a piece of wire by repeated bending.

8. Distinguish between malleability and ductility.

A malleable material may be rolled (or beaten) into sheets (example aluminum). A ductile materials may be drawn into a wire (example - copper and aluminum).

9. What are the units of the shear modulus ?

N/m^2 or lb/in^2 , which are the units for all moduli since stress is F/A and strain is unitless as the ratio of lengths.

10. Why do liquids have only bulk moduli and how are they related to compressibilities ?

Liquids cannot support a tensile strength or shear stress. They have no longitudinal or shear moduli. The compressibility is the reciprocal of the bulk modulus.

Sample Quiz

(Remove the quiz from the book and test your knowledge of the chapter material as though you were taking an in-class quiz. Check your answers with the key at the back of the Study Guide.)

Completion

1. On a stress vs. strain graph, Young's modulus is the
   _____ of the curve in the elastic region.

2. A material will fracture when the _____ is
   exceeded.

3. The ratio of the volume stress and strain is called the
   _____.

Multiple Choice.

___ 4. After which of the following does a material show no recovery to
   its original shape or dimensions ?
   A. elastic deformation   C. elastomeric deformation
   B. plastic deformation   D. liquid volume deformation

___ 5. A material that can withstand only a limited amount of stress is
   said to be
   A. ductile   B. malleable   C. brittle   D. hard

Problems

6. Aluminum has a Young's modulus of 7.0×10^{10} N/m^2. Could an aluminum wire with a diameter of 0.10 cm be stretched to increase its length by 0.50%. The tensile strength for aluminum is 1.4×10^8 N/m^2.

7. A pressure of 200 lb/in^2 produces a fractional change of 0.060 in the volume of a particular liquid. What is compressibility of the liquid ?

Chapter 13   Vibrations and Waves

Sample Problems

Waves and Wave Motion

Example   A force of 0.30 N is applied to a string whose length is
0.30 m and mass 3.0 g.
A. Find the mass density of the string.
B. Find the speed of a wave moving down the string.

Solution

given : $F = 0.30\,N$; $m = 3.0 \times 10^{-3}\,kg$; $L = 0.30\,m$

A. $\mu = m / L$
$\mu = (3.0 \times 10^{-3}) / 0.30$
$\mu = 1.0 \times 10^{-2}\,kg/m$

B. $v = (F / \mu)^{1/2}$
$v = [\,0.30 / (\,1.0 \times 10^{-2})\,]^{1/2}$
$v = 5.5\,m/s$

Periodic Motion

Example 1   Five waves pass a point in 2.0 s. The distance between
the highest point between adjacent waves is 0.80 m.
A. Find the period of the waves.
B. Find the frequency of the waves.
C. Find the speed of the waves.

Solution

given : 5 waves in 2.0 s ; $\lambda = 0.80\,m$

A. $f = 5 / 2 = 2.5\,Hz$

B. $T = 1 / f$
$T = 1 / 2.5$
$T = 0.40\,s$

C. $v = \lambda f$
    $v = (0.80)(0.40)$
    $v = 0.32$ m/s

Example 2 A 5.0 g mass oscillates according to the equation :
           x = 8.0 cos 2t (cm).
           A. Find the amplitude.
           B. What is the angular frequency ?
           C. What is the period ?
           D. What is the frequency ?
           E. Find the total energy of the particle.
           F. Find the maximum speed of the particle.
Solution

           given : A = 8.0 cm ; $\omega = 2.0$ rad/s

A. $A = 8.0$ cm
B. $\omega = 2.0$ rad/s
D. $\omega = 2\pi f$
    $2.0 = 2 (3.14) f$
    $f = 0.32$ Hz
C. $T = 1 / f$
    $T = 1 / 0.32$
    $T = 3.14$ s
E. $E_{max} = (1/2)kA^2$
    $k = \omega^2 m$
    $k = (2.0)^2(5.0 \times 10^{-3})$
    $k = 2.0 \times 10^{-2}$ N/m
    $E_{max} = (1/2)(2.0 \times 10^{-2})(0.08)^2$
    $E_{max} = 6.4 \times 10^{-5}$ J
F. $v_{max} = A\omega$
    $v_{max} = (0.08)(2.0)$
    $v_{max} = 0.16$ m/s

Example 3   Find the length of a pendulum on the earth's surface if the period measures 2.0 s.

Solution

given : $T = 2.0$ s ; $g = 9.8$ m/s^2

$T = 2\pi (L/g)^{1/2}$

$T^2 = 4\pi^2 (L/9.8)$

$2.0^2 = 4.0 L$

$L = 1.0$ m

Wave Characteristics

Example 1   A radio wave travels with a speed of 3.0×10^8 m/s. The wave has a frequency of 101 MHz. Find the wavelength of the wave.

given : $v = 3.0 \times 10^8$ m/s ; $f = 101 \times 10^6$ Hz$=1.01 \times 10^8$ Hz

Solution

$v = \lambda f$

$3.0 \times 10^8 = \lambda (1.01 \times 10^8)$

$\lambda = 3.0$ m

Example 2   A fisherman notices that ten waves pass him in 15 s. He also estimates the distance between waves is 1.0 m. Calculate the speed of the waves.

Solution

given : 10 waves in 15 s ; $\lambda = 1.0$ m

$f = 10 / 15 = 0.67$ Hz

$v = \lambda f$

$v = (1.0)(0.67)$

$v = 0.67$ m/s

Standing Waves

Example 1   A standing wave with four loops is produced in a string 2.0 m long. The speed of the waves is 20 m/s. Find the frequency.

<u>Solution</u>

$$2\lambda = L$$
$$2\lambda = 2.0 \text{ m}$$
$$\lambda = 1.0 \text{ m}$$

$$v = \lambda f$$
$$20 = (1.0) f$$
$$f = 20 \text{ Hz}$$

Example 2   A string 4.0 m long has a mass of 0.20 kg when a force of 5.0 N is applied to the string. When the cord vibrates, it does so as an entire unit. Find the wavelength and the frequency of the wave.

<u>Solution</u>

$$L = 4.0 \text{ m} \;\; ; m = 0.40 \text{ kg} ; F = 5.0 \text{ N}$$
$$\mu = m / L$$
$$\mu = 0.20 / 4.0$$
$$\mu = 0.05 \text{ kg/m}$$
$$v = (F / \mu)^{1/2}$$
$$v = (5.0 / 0.05)^{1/2}$$
$$v = 10 \text{ m/s}$$

$$\lambda = 2L$$
$$\lambda = 2 (4.0) = 8.0 \text{ m}$$
$$v = \lambda f$$
$$10 = (8.0) f$$
$$f = 1.25 \text{ Hz}$$

<u>Solutions to Selected Problems from the Text</u>

5. given: $L = 4.0 \text{ m} \; ; m = ? \; ; v = 15 \text{ m/s} \; ; T = 20 \text{ N} \; ; \mu = \text{mass} / \text{length}$

$$v = (F / \mu)^{1/2}$$
$$v^2 = F / \mu$$

$15^2 / 20 / \mu$
$\mu = 20 / 225$
$\mu = 0.089$ kg/m
$\mu = m / L$
$\mu L = m$
$(0.089)(4.0) = 0.36$ kg

11. given : $f = 5.0$ Hz ; $T_1 = 2.0$ s ; $T_2 = 0.6$ s

frequency is defined as the number of oscillations per unit time.

5 oscillation / 1.0 s = n oscillations / 2.0 s
10 oscillations

5 oscillations / 1.0 s = n oscillations / 0.6 s
3 oscillations

15. given : period = 1.0 s ; $g = 9.8$ m/s^2

$T = 2\pi (L/g)^{1/2}$
$1.0^2 = 4\pi^2 L / (9.8)$
$1.0 = 4.0 L$
$L = 0.25$ m

20. given : $y = 8.0 \sin (10)t$ (cm)

8.0 cm = amplitude
10.0 cm = angular frequency

A. $\omega = 2\pi f$
$10 = 2\pi f$
$10 = 6.28 f$
$f = 1.6$ Hz

B. $T = 1 / f = 1 / 1.6$ Hz $= 0.63$ s
$y = 8.0 \sin 10(0.21)$ cm
$y = 8.0 \sin 2.1$
$y = 8 (0.86)$
$y = 6.9$ cm

25. given : $m = 0.15\,kg$; $k = 150\,N/m$; $y = (0.12\,m)\cos\omega t$

the amplitude of the SHM is 0.12 m

$E = (1/2)kA^2$
$E = (1/2)(150)(0.12)^2$
$E = 1.1\,J$

22. A. Since the clock runs slow , the pendulum should be shortened.

$T = 2\pi (L/g)^{1/2}$
$L/L_0 = (T/T_0)^{1/2}$
$L/L_0 = (60/65)^{1/2}$
$L/L_0 = 0.85$
$\Delta L = L_0 - L$
$\Delta L = L_0 - (0.85)L_0$
$\Delta L = (0.15)(0.35\,m)$
$\Delta L = 0.053\,m = 5.3\,cm$

34. given : T (the period of the wave) = 2.0 s ;
λ (the wavelength) =1.5 m

$f = 1/T = 1/2 = 0.5\,Hz$
$v = \lambda f$
$v = (1.5)(2.0)$
$v = 3.0\,m/s$

B. The wave speed is dependent on the medium in which it is traveling - water. Therefore the speed of the waves does not depend upon the speed of the boat. The amplitude will be affected by the speed of the boat.

39. given : $m = 0.25\,kg$; $L = 5.0\,m$; $T = 40\,N$; fundamental frequency

132

First find the speed of the wave .

$$V^2 = T / \mu$$
$$v^2 = 40 / 0.0.05$$
28.3 m/s

The fundamental would have a single loop and nodes at the ends. The wavelength would equal twice the length or 10.0 m

$$v = \lambda f$$
$$28.3 = (10.0) f$$
$$f = 2.8 \text{ Hz}$$

B.   Harmonics are whole number multiples of the fundamental frequency.

second harmonic will be 5.6 Hz and the third harmonic would be 8.4 Hz.

Review Questions

1. What is a wave ?

It is a disturbance whereby energy is transferred due to a physical disturbance without the transfer of matter. Electromagnetic waves (studied in a later chapter) can propagate through a vacuum.

2. Why will transverse waves propagate in liquids and gases ?

Because liquids and gases cannot support a shear. There is no restoring force in the transverse direction.

3. What is necessary for simple harmonic motion ?

A restoring force that is proportional to the displacement or Hooke's law condition.

4. What determines the amplitude of an object in SHM ?

The initial conditions such as how far it was displaced initially or its initial velocity.

5. Distinguish between frequency and angular frequency in SHM.

The frequency (f) is the number of cycles per second of an oscillation. The angular frequency (ω) is the angular speed of the corresponding circular motion representation of SHM. The angular frequency equal 2π times the frequency.

6. What determines the total energy in SHM ?

The total energy is proportional to the square of the amplitude (A^2), hence, is determined by the initial conditions.

7. A wavelength is the distance between adjacent particles that are in phase. What does "in phase" mean ?

That the particles have the same phase constant of move in unison. The particles have the same displacement with time. If particles are out of phase, they move differently. These particles move in the opposite directions and / or with different speeds.

8. What would be the result of two wave pulses, one with an amplitude A and the other with amplitude 2A, in total constructive interference ? How about total destructive interference ?

By the principle of superposition, the waves would combine to give a pulse with amplitude of 2A for constructive interference, depending on how an amplitude of +A of -A for destructive interference, depending on how the pulses are oriented - (whether 2A is up or down).

9. Distinguish between a node and an antinode.

A node is a position of zero displacement in a standing wave. An antinode is a position of maximum displacement.

10. What is different about resonance for a vibrating string and a mass oscillating on a spring ?

A vibrating string can have any number of discrete resonant frequencies. That is the string will be in resonance when driven at any of its natural of characteristic frequencies. A mass oscillating on a spring, however, has only one oscillation frequency and hence only one resonant frequency.

Sample Quiz

(Remove the quiz from the book and test your knowledge of the chapter material as through you were taking an in-class quiz.   Check your answers with the key at the back of the Study Guide.)

Completion

1. Particle motion in longitudinal waves is_____.

2. In the general equation for SHM, the two quantities determined by the initial conditions are the _____ and the _____ .

3. The characteristic frequency of a string depends on the intrinsic material property of _____ of the string itself.

Multiple Choice.

___ 4. The ratio of v / f for a wave is equal to its
    A. wavelength   C. angular frequency
    B. phase constant   D. none of the choices

___ 5. A standing wave in a string being driven at its second overtone frequency will have how many wavelengths in the string ?
    A. 1   B. 1.5   C. 2   D. 2.5

Problems

6. A particle oscillating in SHM is described by y = 2.0 sin(5.0)t cm.
     A. What is the amplitude ?
     B. What is the frequency of oscillation ?

7. The wave speed in a stretched string with a length of 2.0 m is
    400 m/s.   If the observed standing wave has three nodes, what is the
    harmonic of a value of the frequency of vibration ?

Chapter 14   Sound

Sample Problems

The Nature of Sound

Example 1   Find the speed of sound in glass.

Solution

given : $\rho = 2.6 \times 10^3$ kg/m^3 ; B = 36×10^{10} N/m^2

$v = (B/\rho)^{1/2}$
$v = [\,(36 \times 10^{10})\,/\,(2.6 \times 10^3)\,]^{\,1/2}$
$v = 1.18 \times 10^3$ m/s

Example 2   Find the speed of sound in air on a cold day when the
temperature is
A. 20°F
B. 10°C

Solution

given : temperatures 20°F and 10°C

A. $v = 1087 + 1.1 (T_F - 32°)$ ft/s
   $v = 1087 + 1.1 (20 - 32)$ ft/s
   $v = 1087 + -13$
   $v = 1074$ ft/s

B. $v = 331 + (0.6)T_c$
   $v = 331 + (0.6)(10)$
   $v = 337$ m/s

Sound and Hearing

Example 1　The intensity 2.0 m from a source is 5.0×10^{-10} W/m^2. Find the intensity of the sound if the distance is changed to 10.0 m.

Solution

given : $I_1 = 5.0 \times 10^{-10}$ W/m^2 ; $r_1 = 2.0$ m ; $r_2 = 10.0$ m

$I_1 / I_2 = (r_2 / r_1)^2$

$(1.0 \times 10^{-10}) / I_2 = (10 / 2)^2$

$(1.0 \times 10^{-10}) / I_2 = 25 / 1$

$(1.0 \times 10^{-10}) / I_2 = 25 / 1$

$I_2 = 4.0 \times 10^{-12}$ W/m^2

Example 2　A sound has an intensity of 5.0×10^{-6} W/m^2. Find the intensity in decibels.

Solution

$\beta = 10 \log (I / I_o)$

$\beta = 10 \log [(5.0 \times 10^{-6}) / (1.0 \times 10^{-12})]$

$\beta = 10 \log (5.0 \times 10^{6})$

$\beta = 10 (6.7)$

$\beta = 67$ dB

Example 3　The sound level in dB for a jet plane 300 m away is 105 dB. What is the intensity level of two planes 300 m from the observer.

Solution

DO NOT ADD 105 dB + 105 dB.

$\beta = 10 \log I / I_o$

$105 = 10 \log [I / (1.0 \times 10^{-12})]$

$$10.5 = \log I / (1.0 \times 10^{-12})$$
$$3.16 \times 10^{10} = I / (1.0 \times 10^{-12})$$
$$I = 3.16 \times 10^{-2} \text{ W/m}^2 \qquad \text{the intensity from one plane}$$
$$I = 6.32 \times 10^{-2} \text{ W/m}^2 \qquad \text{the intensity from two planes}$$
$$\beta = 10 \log [(6.32 \times 10^{-2}) / (1.0 \times 10^{-12})]$$
$$\beta = 10 \log 6.32 \times 10^{10}$$
$$\beta = 10 (10.8)$$
$$\beta = 108 \text{ dB}$$

Sound Phenomena

Example 1   An organ pipe open on one end has a length of 85 cm. Find the frequency of the fundamental if the air temperature is
A. 20°C
B. 35°C

Solution

given : $L = 0.85$ m ; $T_1 = 20°C$; $T_2 = 35°C$

$\lambda = 4 L$

$\lambda = 4 (0.85$ m$)$

$\lambda = 3.4$ m

$v_1 = 331 + (0.6)(20) = 343$ m/s ;

$v_2 = 331 + (0.6)(35) = 352$ m/s

$v = \lambda f$

$343 = (3.4) f_1$; $353 = (3.4) f_2$

$f_1 = 100$ Hz $f_2 = 104$ Hz

Example 2   An open organ pipe produces a fundamental frequency of 212 Hz when the air temperature is 20°C. Find the length of the pipe.

Solution

given : $f = 212$ Hz ; $T = 20°C$

$v = 331 + (0.6)(20) = 343$ m/s

$v = \lambda f$

$343 = \lambda \,(212)$

$\lambda = 1.6$ m

$\lambda = 2 L$

$L = 1.6 / 2$

$L = 0.80$ m or 80 cm

Example 3   Two trumpets produce tones which have frequencies of 256 Hz and 259 Hz.   What is the beat frequency ?

Solution

$f_1 = 256$ Hz ; $f_2 = 259$ Hz

beat frequency $= f_2 - f_1$

beat frequency $= 259 - 256 = 3$ beats per second or 3 Hz

Example 4   A person driving toward a stationary sound source (frequency 700 Hz) has a speed of 20 m/s. Find the frequency heard by the driver as the person
A. approaches the source.
B. moves away from the source.

Solution

assume the speed of sound is 340 m/s

A. $f = f_s \,[\,(v \pm v_o)/v\,]$

$f = (700) \,[\,(340 + 20)/340\,]$

$f = 700 \,(1.06)$

$f = 741$ Hz

B. $f = (700) \,[\,(340 - 20)/340\,]$

$f = (700)\,(0.94)$

$f = 659$ Hz

Example 5   A train approaches a stationary observer. The engineer hears a 300 Hz pitch and the observer hears a frequency of 315 Hz. Assuming the speed of sound is 340 m/s, find the speed of the train.

141

Solution

$$f = f_s [v / (v \pm v_s)]$$
$$315 = 300 [340 / (340 \pm v_s)]$$
$$1.05 = 340 / 340 \pm v_s$$
$$340 \pm v_s = 329$$
$$v_s = 11 \text{ m/s toward the observer}$$

Solutions to Selected Problems from the Text

4. given : Y for copper (Young's modulus) $= 11 \times 10^{10}$ N/m^2
    density $= 8.9 \times 10^3$ kg/m^3

$$v = (Y / \rho)^{1/2}$$
$$v = (11 \times 10^{10} / 8.9 \times 10^3)^{1/2}$$
$$v = 3.5 \times 10^3 \text{ m/s}$$

9. given : B for mercury (Bulk modulus) $= 2.8 \times 10^{10}$ N/m^2
    density for mercury $= 13.6 \times 10^3$ kg/m^2

$$v = (B / \rho)^{1/2}$$
$$v = (2.8 \times 10^{10} / 13.6 \times 10^3)^{1/2}$$
$$v = 1.4 \times 10^3 \text{ m/s}$$

14. given : $\Delta v = 10$ m/s $\quad ; \quad \Delta v = 10$ ft/s

   A. $\Delta T_C = v / 0.6 = 10 / 0.6 = 16.7 \ °C$

   B. $\Delta T_F = v / 1.1 = 10 / 1.1 = 9.1°F$

18. given : $f = 50 \times 10^6$ Hz ; information to find the speed of sound in alcohol

First find the speed of sound in alcohol

$$v = (B / \rho)^{1/2}$$
$$v = (0.11 \times 10^{10} / 0.79 \times 10^3)^{1/2}$$
$$v = 1.18 \times 10^3 \text{ m/s}$$

Now find the wavelength.

$$v = \lambda f$$
$$1.18 \times 10^3 = \lambda \ (50 \times 10^6)$$
$$\lambda = 2.4 \times 10^{-5} \text{ m}$$

22. given : $I = 5.0 \times 10^{-7}$ W/m^2 ; $r_1 = 1.5$ m ; $r_2 = 2.5$ m

$$I_1 / r_1^2 = I_2 / r_2^2$$
$$(5.0 \times 10^{-7}) / 1.5^2 = I_2 / 2.5^2$$
$$I_2 = 1.8 \times 10^{-7} \text{ W/m}^2$$

27. given : $I_1 = 1 \times 10^{-2}$ W/m^2 ; $I_2 = 1.0 \times 10^{-6}$ W/m^2

$$IL_1 = 10 \log (I / I_o)$$
$$IL_1 = 10 \log [(1 \times 10^{-2}) / (1 \times 10^{-12})] = 100 \text{ dB}$$

$$IL_2 = 10 \log (I / I_o)$$
$$IL_2 = 10 \log [(1.0 \times 10^{-6}) / (1.0 \times 10^{-12})] = 60 \text{ dB}$$
$$IL_2 - IL_1 = 20 \text{ dB} - 60 \text{ dB} = -40 \text{ dB}$$

35. given : temperature = 20 °C ; $f = 256$ Hz

$$v = 331.5 \text{ m/s} + (0.6)(20) = 343 \text{ m/s}$$
$$v = \lambda f$$
$$343 = \lambda(256)$$

143

$$\lambda = 1.34 \text{ m}$$
$$\lambda = 4L$$
$$L = 1.34 / 4 = 0.33 \text{ m}$$

41.   given : $f = 500 \text{ Hz}$; $f' = 520 \text{ Hz}$; $v_{sound} = 340 \text{ m/s}$

$$f' = f[v/(v - v_s)]$$
$$520 = 500[340/(340 - v_s)]$$
$$1.04 = 340/(340 - v_s)$$
$$340 - v_s = 327$$
$$v_s = 340 - 327 = 13 \text{ m/s}$$

Review Questions

1. What is the frequency range of the human ear ?

    20 Hz to 20,000 Hz

2. If you plotted the speed of sound v versus the Celsius temperature, T_C, describe the graph you would obtain.

    A straight line with a slope of 0.60 and a y-intercept of 330 m/s.

3. How many dB increase in the intensity level would increase the intensity by a factor of 2 and a factor 100 ?

    To double the intensity (factor of 2) requires a 3 dB increase, since $10 \log(I/I_o) = 10 \log(2) = 10(0.3) = 30 \text{ dB}$.   For 10 fold increases, the increase in dB is ten times the exponent of 10 equal to the factor, which is the
    increase in bels, $10^2 = 100$, and $2 \times 10 = 20 \text{ dB}$.

4. How does sound quality depend on wave form ?

Quality is associated with the number of overtones in a sound,
which affects thee wave form by the principle of superposition.

5. What produces beats ?

It depends on the wave energy or amplitude and also the
frequency. Some sounds of different intensity and different
frequencies are judged to be equally loud.

6. On what does perceived loudness depend ?

It depends on the wave energy or amplitude and also the
frequency. Some sounds of different intensities and different
frequencies are judged to be
equally loud.

7. What are the boundary conditions for forming standing waves in a
closed pipe ?

Since a closed pipe has one open end, there must be a node at
the closed end and an antinode at the open end.

8. What would be the affect if you were traveling at one speed and a
sound source followed you at a greater speed ?

There would be a Doppler shift in sound source frequency just
as though a moving source approached a stationary observer
with a speed equal to the relative speed when both were
moving.

9. Distinguish between ultrasonic and supersonic.

Ultrasonic refers to sound with frequencies above the human
audible range, that is, greater than 20 kHz. Supersonic refers to
something traveling faster than the speed of sound.

145

10. What is the principle of an ultrasonic cleaning bath ?

Ultrasonic with relative high frequencies has relatively short wavelengths which are on the order of small cavities, which help dislodge particles of foreign matter.

Sample Quiz

(Remove the quiz from the book and test your knowledge of the chapter material as through you were taking an in-class quiz. Check your answers with the key at the back of the Study Guide.)

Completion

1. Since the speed of sound in matter depends on material moduli, the speed of sound is generally greatest in _____.

2. The threshold of human hearing has an intensity level of _____dB.

3. When a sound source pulls away from a moving observer, a_____ in frequency is observed.

Multiple Choice.

____ 4. The approximate speed of sound in air is
   A. 1.5 mi/s   B. 1/2 mi/s   C. 1.2 mi/s   D. none of the choices

____5. In going from the intensity level of a whisper to a loud radio, there is an increases of 50 dB. This corresponds to an intensity increase of
   A. 50   B. 10^4 C. 5×10^4 D. 10^5

Problems

6. The intensity of a sound is decreased from 10^{-4} W/m^2 to 10^{-7} W/m^2. What is the\ corresponding decrease in the intensity level ?

7. A sound source with a frequency of 400 Hz approaches a stationary observer wit speed of 25 m/s on a hot day when the temperature is 38°C. What is the frequency heard by the observer ?

8. An organ pipe is open on one end and is 20 cm long. If the speed of sound is 340 m/s, find the frequency of the fundamental and first audible harmonic.

Chapter 15   Fluid Mechanics

Sample Problems

Fluid Properties

Example 1   A room has the dimensions of 12 ft x 16 ft x 9 ft. Find the weight of the air in the room.

Solution

given : dimensions to calculate the volume ;

D_{air} = 0.0807 lb/ft^3

V = (12 ft)(16 ft)(9 ft)

V = 1728 ft^3

D = w / V
w = D V

w = (0.0807)(1728 ft^3)
w = 139 lb

Example 2   Five hundred grams of water are added to 200 g of sea water. Find the specific gravity for the 700 g mixture.

Solution

given : 500 g of H_2O ; 200 g of sea water

ρ = m / V
1.00 = 200 / V

V_1 = 200 cm^3

1.03 = 500 / V

V_2 = 485 cm^3

ρ = 700 / 685
ρ = 1.02 g/cm^3

Pressure and Pressure Measurement

Example 1   A pool is 3.0 m deep. Find the gauge and absolute pressure at the bottom of the pool.

Solution

$$\rho_{water} = 1.0 \times 10^3 \text{ kg/m}^3 \; ; \; h = 3.0 \text{ m}$$
$$\Delta P = \rho g h$$
$$\Delta P = (1.0 \times 10^3)(9.8)(3.0)$$
$$\Delta P = 2.94 \times 10^4 \text{ N / m}^2 \quad \text{- gauge pressure}$$
$$P_{absoute} = 1.013 \times 10^5 \text{ N/m}^2 + 2.94 \times 10^4 \text{ N/m}^2$$
$$P_{absolute} = 1.31 \times 10^5 \text{ N/m}^2$$

Example 2   A person drives into the mountains changing the elevation by 1000 m. Calculate the force which would be exerted on the ear drum if the drum has an area of 0.50 cm^2.

Solution

$$\text{given : } \rho_{air} = 1.29 \text{ kg/m}^3 \; ; \; \Delta h = 1000 \text{ m} \; ;$$
$$A = 0.50 \text{ cm}^2 = 5.0 \times 10^{-5} \text{ m}^2$$
$$\Delta P = \rho_{air} g h$$
$$\Delta P = (1.29)(9.8)(1000) = 1.26 \times 10^4 \text{ N/m}^2$$
$$\Delta P = F / A$$
$$F = \Delta P A$$
$$F = (1.26 \times 10^4)(5.0 \times 10^{-5})$$
$$F = 0.63 \text{ N}$$

Buoyancy and Archimede's Principle

Example 1   A piece of lead has a mass of 500 g in air. Find the weight of the lead in the water.

Solution

given : $m = 0.50$ kg ; $\rho_{lead} = 11.4 \times 10^3$ kg/m^3 ;

$\rho_{water} = 1.0 \times 10^3$ kg/m^3

$\rho = m / V$

$11.4 \times 10^3 = 0.500 / V$

$V = 4.4 \times 10^{-5}$ m^3

$F_b = \rho_{water} V g$

$F_b = (1.0 \times 10^3)(4.4 \times 10^{-5})(9.8) = 0.43$ N

$F = mg - F_b$

$F = (0.5)(9.8) - 0.43$

$F = 4.5$ N

Example 2   A rectangular wooden block has a mass of 250 g. Find the minimum force required to keep the block submerged.

Solution

given : $\rho_{wood} = 810$ kg/m^3 ; $m = 0.25$ kg ;

$\rho_{water} = 1.0 \times 10^3$ kg/m^3

$\rho = m / V$

$810 = 0.25 / V$

$V = 3.1 \times 10^{-4}$ m^3

$F_b = \rho V g$

$F_b = (1.0 \times 10^3)(3.1 \times 10^{-4})(9.8)$

$F_b = 3.0$ N

$w = mg = (0.25)(9.8) = 2.5$ N

$F_{downward} = 0.5$ N

Example 3   A 20 cm^3 block has an apparent mass of 5.0 g in water.
A. Find the mass of the block.
B. Find the apparent mass of the block in alcohol.

Solution

A. given : $V = 20 \text{ cm}^3$; $m_{app} = 5.0$; $\rho_{water} = 1.0 \text{ g}/\text{cm}^3$

$mg_{app} = mg - F_b$

$mg_{app} = mg - \rho_{water} Vg$

$5.0 \quad = m - (1.0)(20)$

$m \quad = 25 \text{ g}$

B. $m_{app} = 25 - (0.79)(20)$

$m_{app} = 25 - 16$

$m_{app} = 16 \text{ g}$

Fluid Flow

Example 1   Water flows in a horizontal pipe with varying cross-section area. When the diameter of the pipe is 20 cm, water flows with a speed of 10 m/s.
A. Find the speed of the water in the pipe if the diameter of 50 cm.
B. Find the difference in pressure between the two cross sections.

Solution

given : $v_1 = 1.0 \text{ m/s}$; $r_1 = 10 \text{ cm}$; $r_2 = 25 \text{ cm}$

A. the volume rate of flow is the same

$A_1 v_1 = A_2 v_2$

$\pi(10)^2 (10) = \pi(25)^2 v_2$

$v_2 = 1.6 \text{ m/s}$

B. $(1/2)\rho v_1^2 + P_1 = (1/2)\rho v_2^2 + P_2$

$\Delta P = (1/2)\rho (v_1^2 - v_2^2)$

$\Delta P = (1/2)(1000)(10^2 - 1.6^2)$

$\Delta P = 4.87 \times 10^4 \text{ N/m}^2$

Example 2 Air rushes on the roof of a home during a wind storm with a speed of 50 m/s. Find the force on a roof whose area is 100 m^2.

Solution

given : $\rho_{air} = 1.29$ kg/m^3 ; A = 100 m^2 ; v = 50 m/s

$$\Delta P = (1/2)\rho_{air} \, v^2$$

$$\Delta P = (1/2) \, (1.29)(50)^2$$

$$\Delta P = 1.6 \times 10^3 \text{ N/m}^2$$

$$P = F / A$$

$$F = P \, A$$

$$F = (1.6 \times 10^3)(100)$$

$$F = 1.6 \times 10^5 \text{ N}$$

Example 3 Water flows at 20 m/s through a 50 cm radius pipe into the basement of a building. The pipe tapers to a radius of 20 cm at a height of 10 m above the basement.
A. Calculate the speed of the water at the new height.
B. Find the difference in pressure between the two heights.

Solution

given : $v_1 = 20$ m/s ; $r_1 = 50$ cm ; $r_2 = 30$ cm ;

$\rho_{water} = 1.0 \times 10^3$ kg/m^3

A. $$A_1 v_1 = A_2 v_2$$

$$\pi \, (50)^2 \, (20) = \pi \, (30)^2 v_2$$

$$v_2 = 55 \text{ m/s}$$

B. $$P_1 + \rho g h_1 + (1/2)\rho v_1^2 = P_2 + \rho g h_2 + (1/2)\rho v_2^2$$

$$P_1 + 0 + (1/2)(1.0 \times 10^3)(20)^2 =$$

$$P_2 + (1.0 \times 10^3)(9.8)(10) + (1/2)(1.0 \times 10^3)(55)^2$$

$$\Delta P = P_1 - P_2 = 1.41 \times 10^6 \text{ N/m}^2$$

Solutions to Selected Problems from the Text

5. given: $V_{alcohol} = 50 \text{ mL}$; $V_{gasoline} = 50 \text{ mL}$;

$\rho_{alcohol} = 0.79 \text{ g/cm}^3$; $\rho_{gasoline} = 0.68 \text{ g/cm}^3$

$$m_{alcohol} + m_{gas} = m_{total}$$
$$\rho V_{alcohol} + \rho V_{gas} = mass_{total}$$
$$(0.79)(50) + (0.68)(50) = mass_{total}$$
$$73.5 \text{ g} = m_{total}$$

10. given : radius of tube $= 0.5 \times 10^{-3}$ cm ; $h = 8 \times 10^{-3}$ m

a. $F = mg = \rho_{water} V g$

$F = (1000)\pi(0.5 \times 10^{-3})^2(8.0 \times 10^{-3})\,(9.8)$

$F = 6.2 \times 10^{-2}$ N

b. $h_b = (\rho_{water} / \rho_{blood})(h_w)$

$h_b = (1.0 / 1.1)(0.80) = 0.73$ cm

14. given : density of water $= 1.0 \times 10^3 \text{ kg/m}^3$; depth $= 2.5$ m ;

$P_{atmospheric} = 1.013 \times 10^5 \text{ N/m}^2$

$P_{absolute} = P_g + P_{atmospheric}$

$P_{absolute} = (1.0 \times 10^3)(9.8)(2.5) + 1.013 \times 10^5$

$P_{absolute} = 1.26 \times 10^5 \text{ N/m}^2$

20. given : height difference $= 15$ cm ;

density of mercury $= 13.6 \text{ g/cm}^3$

$P_g = \rho_{mercury} V g$

$P_g = (13.6 \times 10^3)\,(9.80)\,(0.15) = 2.0 \times 10^4 \text{ N/m}^2$

23. given : $p_g = 100 \text{ N/m}^2$; $\rho_{mercury} = 13.6 \times 10^3 \text{ kg/m}^3$

open : $P_g = \rho g h$

$$100 = (13.6 \times 10^3)(9.8)(h)$$
$$h = 7.5 \times 10^{-4} \text{ m}$$

closed : $P = \rho g h$

$$(1.013 \times 10^5 + 100) = (13.6 \times 10^3)(9.8) h$$
$$h = 0.76 \text{ m or 76 cm}$$

28. given : $V = 0.20 \text{ ft}^3$; $w = 20 \text{ lb}$

$$F_B = w - w_{ap}$$
$$w_{ap} = w - F_b$$
$$w_{ap} = 20 - (0.20)(55)$$
$$w_{ap} = 9.0 \text{ lb}$$

34. given : diameter = 1.0 in (radius = 0.5 in) ; $v = 1.2 \text{ ft/s}$;
$V = 100 \text{ gal}$

$$V = (100 \text{ gal}) \,(0.134 \text{ ft}^3/\text{gal}) = 13.4 \text{ ft}^3$$

$$av = V / t$$
$$\pi \, (4.2 \times 10^{-2})^2 (1.2) = 13.4 / t$$
$$6.65 \times 10^{-3} = 13.4 / t$$
$$t = 13.4 / (6.65 \times 10^{-3}) = 2.0 \times 10^3 \text{ s} = 34 \text{ min}$$

38. given : $h = 10 \text{ m}$

(a) $(1/2)\rho v^2 = \rho g h$

$$v = (2gh)^{1/2}$$
$$v = [(2)(9.8)(10)]^{1/2}$$
$$v = 14 \text{ m/s}$$

(b) $\quad Q = Av$

$$Q = (0.20 \text{ cm}^2) \, (1 \text{ m} / 100 \text{ cm})^2 \, (14 \text{ m/s})$$
$$Q = 2.8 \times 10^{-4} \text{ m}^3/\text{s}$$

43. $\quad$ given : $r_2 = 0.98 \, r_1$

(a) $\quad Q = \pi r^4 \Delta P / 8nL$
the only terms which change are ΔP and r therefore set up a proportion.

$$\Delta P_1 / \Delta P_2 = (r_1 / r_2)^4$$
$$\Delta P_1 / \Delta P_2 = (r_1 / 0.98 \, r_1)^4$$
$$\Delta P_1 / \Delta P_2 = 0.926$$
$$\Delta P_2 = (1.08) \, \Delta P_1$$

(b) $\quad$ Pressure increase of 8% which would cause the heart to work harder.

Review Questions

1. How is weight density related to mass density ?

$D = \rho g$, since $D = w / V = mg / V = (m / V)g = \rho g$

2. Is capillary action based on cohesive or adhesive forces ?

Both, since a liquid must wet the capillary (adhesive forces) and surface tension (cohesive forces) is required for column height change.

3. Distinguish between absolute pressure and gauge pressure.

   Absolute pressure is measured relative to absolute zero pressure or complete vacuum. Gauge pressure is measured relative to atmospheric pressure, and $P_{abs} = P_{gauge} + P_{atm}$

4. What does a negative gauge pressure indicate ?

   A negative gauge pressure indicates a pressure lower than atmospheric pressure or partial vacuum.

5. How does the pressure-depth equation indicate the presence of a buoyant force on submerged objects ?

   Since $P = \rho gh$, the pressure on the bottom surface (Pascal's principle) at a greater depth is greater than the pressure on the top of the object at a lesser depth. As a result, there is a net upward or buoyant force.

6. If a liquid is compressible, what effect would this have on an object sinking in the liquid ?

   If a liquid is compressible, its density will increase with depth. Hence, a sinking object would sink until it reached a depth at which its density equaled the average density of the liquid.

7. What is the advantage of using idealized streamlines in analyzing fluid flow ?

   In streamline flow the particles paths or streamlines never cross, which eliminates whirlpools and eddies, the motion of which is very difficult to analyze.

8. Under what condition would the flow rate of a liquid not be constant ?

If the liquid were compressible, then the mass per unit volume would change and the continuity equation would not apply since the mass through a cross-section would not be conserved or constant.

9. What is the principle of Bernoulli's equation ?

The work-energy theorem or conservation of energy.

10. According to Poiseuille's law, how does the flow rate of a fluid vary with viscosity and the radius of the conduit ?

As would be expected for viscosity, the greater the viscosity, the smaller the flow rate. As might not be expected, the flow rate is directly proportional to r^4, which means if the radius of the conductor is doubled, the flow rate increases by a factor of 16.

Sample Quiz

(Remove the quiz from the book and test your knowledge of the chapter material as through you were taking an in-class quiz.   Check your answers with the key at the back of the Study Guide.)

Completion

1. Droplets of water bead up on a freshly waxed car due to a lack of
   _____ forces and the effect of _____.

2. A barometer used to measure atmospheric pressure measures
    pressure relative to _____.

3. An object will sink in a fluid if the fluid's density is
   _____ than that of the object.

Multiple Choice.

___4. Atmospheric pressure
   A. has a negative absolute pressure
   B. is equal to zero absolute pressure
   C. is equal to zero gauge pressure
   D. none of the above

___5. The flow rates of fluids are commonly measured
   A. using the pressure-depth equation
   B. with a barometer
   C. using Archimedes' principle
   D. with a Venturi meter

Problems

6. A cube of material has a 500 g and is 10 cm on a side. If the top of the cube is placed 50 cm below the surface of a body of water,
    A. what is the pressure difference on the top and bottom of the cube ?
    B. what is the buoyant force acting on the cube ?

7. An ideal liquid flows through a horizontal pipe and then down an incline so the elevation is decreased by one-half and the cross-sectional area of the pipe increases by one-half. What is the pressure difference in terms of the potential and kinetic energies (per unit volume) ?

Chapter 16 Temperature and Heat

Sample Problems

Temperature Measurement and Heat Units

Example 1   Room temperature is said to be 20°C. Find the equivalent temperature on the Fahrenheit, Kelvin, and Rankin scales.

Solution

given : $T_C = 20°C$

$T_F = (9/5)T_C + 32°$
$T_F = (9/5)(20) + 32$
$T_F = 36 + 32$
$T_F = 68°F$

$T_K = T_C + 273$
$T_K = 20 + 273$
$T_K = 293 \text{ K}$

$T_R = T_F + 460$
$T_R = 68 + 460$
$T_R = 528 \text{ °R}$

Example 2   The temperature on a very cold day is -10°F. Find the temperature in Celsius.

Solution

given : $T_F = -10°F$
$T_C = (5/9)(T_F - 32)$
$T_C = (5/9)(-10 - 32)$
$T_C = (5/9)(-42)$
$T_C = -23°C$

161

Example 3   A 150 g baseball falls from a height of 44 m and strikes the ground with a speed of 20 m/s. If the energy lost goes into heat, find the heat generated.

Solution

given : $m = 0.15$ kg ; $h = 44$ m ; $v_{final} = 20$ m/s

$PE = mgh$
$PE = (0.15)(9.8)(44) = 65$ J
$KE = (1/2)mv^2$
$KE = (1/2)(0.15)(20)^2$
$KE = 30$ J
$Q = 35$ J

Example 4   A machine is 60% efficient. It it takes in 1.0×10^3 Btu, find the heat generated.

Solution

given : $eff = 0.60$; $Q_{in} = 1.0 \times 10^3$ Btu ; $Q_{lost} = ?$

$eff = W / Q_{in}$

$0.60 = W / 1.0 \times 10^3$
$W = 600$ Btu
$Q_{lost} = 1000 - 600 = 400$ Btu

Kinetic Theory and the Perfect Gas Law

Example 1   The volume of a has at standard temperature is 5.0 L. The pressure remains constant as the temperature increases to 100°C. Find the new volume.

Solution

given : $T_1 = 273$ K ; $V_1 = 5.0$ L ; $T_2 = 373$ K ; $V_2 = ?$

$V_1 / V_2 = T_1 / T_2$
$5.0 / V_2 = 273 / 373$

$$5.0 / V_2 = 0.73$$
$$V_2 = 6.8 \text{ L}$$

Example 2   A certain gas has a volume of of 20 L at a pressure of 2.0 Atm and a temperature of 10°C. Find the volume of the gas at a temperature of 50°C and a pressure of 2.5 Atm.

Solution

given : $V_1 = 20$ L ; $P_1 = 2.0$ Atm ; $T_1 = 283$ K

$V_2 = ?$; $P_2 = 2.5$ Atm ; $T_2 = 323$ K

$$P_1 V_1 / P_2 V_2 = T_1 / T_2$$
$$(2.0)(20) / (2.5)V_2 = 283 / 323$$
$$1.6 / V_2 = 0.88$$
$$V_2 = 1.8 \text{ L}$$

Specific Heat

Example 1   Fifty calories of heat are added to a 100 g piece of aluminum at a temperature of 10°C. Find the new temperature.

Solution

given : $c_{al} = 0.22$ cal / g-°C ; $m = 100$ g ; $T_0 = 10°C$

$$Q = mc\Delta T$$
$$50 = (100)(0.22)\Delta T$$
$$50 = 22 \ \Delta T$$
$$\Delta T = 2.3°C \qquad ; \ T_{final} = 12.3°C$$

Example 2   A 0.50 kg aluminum contains 1.0 kg of water at 20°C. Find the change in temperature if 10 kcal of heat are added.

Solution

given : $m_{al} = 0.50$ kg ; $c_{al} = 0.22$ kcal / kg-°C ;

$c_w = 1.0$ kcal/kg-°C ;

$\Delta Q = 10$ kcal

$\Delta Q = mc\Delta T_{al} + mc\Delta T_{water}$

$10 = (0.50)(0.22)\Delta T_{al} + (1.0)(1.0)\Delta T_w$

$10 = 0.11 \Delta T + 1.0 \Delta T$

$10 = 1.11 \Delta T$

$\Delta T = 9.0$°C

$T_{final} = 29$°C

Example 3   A 40 g piece of aluminum is heated to 150°C. The aluminum is then placed into 40 g of water at 20°C. Find the new equilibrium temperature.

Solution

given : $m_{al} = 40$ g ; $T_{al} = 150$°C ; $m_w = 40$ g ; $T_w = 20$°C ;

$c_w = 1.0$ cal / g - °C ; $c_{al} = 0.22$ cal / g -°C

$Q_{lost} = G_{gained}$

$mc\Delta T = mc\Delta T$ (mass are equal)

$(0.22) (150 - T) = (1.0) (T - 20)$

$33 - 0.22 T = T - 20$

$53 = 1.22 T$

$T = 43$°C

Phase Changes and Latent Heats

Example 1   How much steam could be condensed of 1.0×10^3 kcal of heat are removed ?

Solution

given : $Q = -1.0 \times 10^3$ kcal ; $L_v = 540$ kcal / kg

$Q = mL_v$

$1.0 \times 10^3 = m (540)$

$m = 1.85$ kg

Example 2 A container holds 400 g of water at 20°C. How much ice would melt to bring the water to a temperature of 0°C ?

Solution

given : mass of water = 400 g ; ΔT_{water} = 20°C ;

L_v = 80 cal / g ; c_{water} = 1.0 cal/g-°C

$Q = mc\Delta T$

      (heat to remove to change water from 20°C to 0°C)

$Q = (0.40)(1.0)(20) = 8.0$ kcal

$Q = mL_f$

$8.0 = m\ (80)$

$m = 0.10$ kg

Example 3 Find the amount of heat needed to change 100 g of ice at -20°C to steam at 120°C.

Solution

given : m = 100 g ; c_{ice} = -.5 cal/g-°C ; L_f = 80 cal/g ;

c_{water}=1.0 cal / g-°C ; L_v = 540 cal/g ;

c_{steam} = 0.48 cal / g - °C

$Q_1 = mc\Delta T$ (heat ice from -20°C to 0°C)

$Q_1 = (100)(0.5)(20) = 1.0$ kcal

$Q_2 = mL_f$ (melt ice)

$Q_2 = (100)(80) = 8.0$ kcal

$Q_3 = mc\Delta T$ (heat water from 0°C to 100°C)

$Q_3 = (100)(1.0)(100) = 10$ kcal

$Q_4 = mL_v$ (change water to steam)

$Q_4 = (100)(540) = 5.4 \times 10^4$ cal = 54 kcal

$Q_5 = mc\Delta T$ (heat stem from 100°C to 120°C)

$Q_5 = (100)(0.48)(20)$

$Q_5 = 960$ cal = 0.96 kcal

$Q_t = 74$ kcal

Solutions to Selected Problems from the Text

3. given : temperatures (a) 43°C   (b) -15°C
                            (c) 50°F   (d) 450 °R

(a) $T_K = T_C + 273$
       $T_K = 43°C + 273°C = 316 K$

(b) $T_K = -15°C + 273°C = 258 K$

(c) $T_C = (5/9) (T_F - 32)$
       $T_C = (5/9) (50 - 32) = 10°C$
       $T_K = 10°C + 273°C = 283°C$

(d) $T_F = T_R - 460$
       $T_F = 450 - 460 = -10°F$
       $T_C = (5/9) (-10 - 32) = -23°C$
       $T_K = -23°C + 273°C = 250 K$

8. given : eff = 0.60 ; $W_{input} = 2000 J$

eff = W_{output} / W_{input}
$0.60 = W_{out} / 2000$
$W_{out} = 1200 J$
$W_{in} = W_{out} + W_{heat}$
$2000 J = 1200 J + W_{heat}$
$W_{heat} = (800 J) (1 kcal / 4180 J) = 0.19 kcal$

14. given : $P_2 = 2P_1$; $V_2 = (0.5)V_1$

$PV = NkT$
$P_1V_1 / P_2V_2 = T_1 / T_2$
$P_1V_1 / 2P_1(0.5V_1) = T_1 / T_2$
$T_1 = T_2$

19. given : $P_1 = 4.0$ L ; $T_1 = 20°C = 293$ K ; $V_1 = 1.6 \times 10^5$ Pa

(a) $P_1V_1 = P_2V_2$
$(4.0)(1.6 \times 10^5) = (1.2 \times 10^5) V_2$
$V_2 = 5.3$ L

(b) $P_1V_1 / P_2V_2 = T_1 / T_2$
$(4.0)(1.6 \times 10^5) / (1.2 \times 10^5)(1.5) = 293 / T_2$
$3.56 = 293 / T_2$
$T_2 = 293 / 3.56$
$T_2 = 82$ K or $-191°C$

24. Aluminum would absorb more heat since it has a greater specific heat.

$\Delta Q_{Al} = mc\Delta T$
$\Delta Q_{Al} = (0.50)(0.21)(32) = 3.4$ Btu
$\Delta Q_{Fe} = (0.50)(0.113)(32) = 1.8$ Btu
the difference will be 1.6 Btu

29. given : $m_{Al} = 250$ g ; $c_{Al} = 0.22$ cal/g-°C ; $m_w = 200$ g ;
$c_w = 1.0$ cal/g-°C

$Q_{lost} = G_{gained}$
$m_{Al}c_{Al}\Delta T_{Al} = m_w c_w \Delta T_w$

$(250)(0.22)(100 - T_f) = (200)(1.0)(T_f - 20)$
$55(100 - T_f) = 200(T_f - 20)$
$5500 - 55T_f = 200 T_f - 4000$
$9500 = 255 T_f$
$T_f = 37°C$

33.   given : $m_m = 500$ g ; $T_{o(metal)} = 100°C$; $m_{water} = 500$ g ;
         $m_{cup} = 250$ g ;
         T_o(cup and water) $= 20°C$; $T_f = 25°C$

         $Q_{lost} = Q_{gained}$
         $(500)(c_m)(\Delta T) = (500)(1.0)\Delta T + (250)(0.22)(\Delta T)$
         $(500)(c_m)(75) = (500)(1.0)(5) + (250)(0.22)(5)$
         $37500(c_m) = 2775$
         $c_m = 0.074$ cal / g-C°

37.   given : initially 10 g of ice at -10°C ; final state is 10 g of water
         at 20°C

         $\Delta Q = \Delta Q_{warm\ ice} + \Delta Q_{melt\ ice} + \Delta Q_{warm\ water}$
         $\Delta Q = (10)(0.50)(10) + (10)(80) + (10)(1.0)(20)$
         $\Delta Q = 1050$ cal

42.   given : $m_{ice} = 40$ g ; $T_{ice} = 0°C$; $m_{water} = 200$ g ;
         $T_{water} = 20°C$

         $Q_{lost} = Q_{gained}$
         $mL + m_w c_w \Delta T_w = m_w c_w \Delta T_w$
         $(40)(80) + (40)(1.0)(T_f - 0) = (200)(1.0)(20 - T_f)$
         $3200 + 40T_f = 4000 - 200 T_f$
         $240 T_f = 4000 - 3200$
         $T_f = 3.3°C$

Review Questions

1. What is temperature a relative measure ?

     Temperature only indicates different degrees of hotness and

coldness and does not be itself indicate the absolute heat content of an object, only the relative content.

2. Can any points be used to define a temperature scale ?

Yes, as long as they are fixed. The ice point and the steam point of water are used because of convenience, Actually, only one point is needed along with a defined by the triple point (a fixed point) of water and the Kelvin interval.

3. What are the advantages of absolute temperature scales ?

Starting at absolute zero, there are no negative temperatures on the absolute scales. Also, the absolute temperature is directly proportional to the internal energy of the internal energy of an ideal gas, which is an important theoretical model.

4. Distinguish between total energy, internal energy, and thermal energy.

The total energy and the internal energy of a substance are the same, i.e., the kinetic and potential energies contained in a body. The thermal energy is the random translational kinetic energy of molecules and is the "temperature" energy that gives the temperature reading.

5. Distinguish between heat capacity and specific heat capacity.

Heat capacity is the ratio of heat and corresponding temperature change of a body. Hence, each object has its own heat capacity, even if two objects are made out of the same substance. Specific heat capacity, or specific heat, is the heat capacity per unit mass (C/m). By taking mass into account each substance has a particular or specific heat capacity.

6. Why is the specific heat of water 1.0 cal/g-°C or 1.0 kcal/kg-°C ?

   Latent heat is associated with a phase change, and at such
   temperatures, the energy or heat goes into work of the phase
   change and not the thermal energy.

7. Why doesn't latent heat go into changing the temperature of a
   substance ?

   Latent heat is associated with a phase change, and at such
   temperatures, the energy or heat goes into the work of the
   phase change and not the thermal energy.

8. Why is the latent heat of vaporization generally greater than the heat
   of fusion for substances ?

   Because it takes more work to separate the molecules in going
   from a liquid to a gas that is it does in going from a solid to a
   liquid.

9. What are the slopes of the slanted lines in Fig 16.13 in terms of
   specific heats ?

   Since $Q = mc\Delta t$ or $\Delta T = (1/mc)Q$, which is of the form $y = mx + b$
   for a straight line, the slopes are equal to $1/mc$.

10. How are the boiling point and freezing point affected by pressure ?

   In general the increase in pressure always favors the liquid state
   of matter. Increased pressure raises the boiling point and
   lowers the freezing point.

Sample Quiz

(Remove the quiz from the book and test your knowledge of the chapter material as through you were taking an in-class quiz. Check your answers with the key at the back of the Study Guide.)

Completion

1. Room temperature on the Celsius scale is taken to be _____ and on the Fahrenheit scale _____.

2. The procedure used to measure the specific heats of material in calorimetry is called the _____.

3. The temperature and pressure at which a substance can exist in all three phases of matter is called the _____.

Multiple Choice

_____ 4. The addition or removable of heat from a body is a given phase changes its
    A. thermal energy   C. total energy
    B. internal energy   D. all of the choices

_____ 5. A substance with a small specific heat
    A. will have a low boiling point
    B. requires a small amount of heat for a relatively large temperature change
    C. will have large latents heats
    D. none of the above

Problems

6. An ideal gas is heated from -10°C to 200°C. By what factor is the internal energy of the gas increased ?

7. One hundred grams of alcohol at room temperature is heated until it is completely vaporized.   If the boiling point of an alcohol is 78°C, how much energy is required to do this ? (c = 0.60 cal/g-°C and L_v = 204 cal/g)

Chapter 17

Thermal Properties of Matter

Sample Problems

Heat Transfer

Example 1 The area for a window 1.0 cm thick is 1.5 m^2. Find the rate of heat flow through the window of the outside temperature is 0°C and the indoor temperature is 20°C.

Solution

given : $k_{glass} = 2.0 \times 10^{-3}$ cal /cm-s-°C ; L = 1.0 cm ;

$A = 1.5 \times 10^4$ cm^2 ; $\Delta T = 20°C$

$\Delta Q / \Delta t = kA \, \Delta T / L$

$\Delta Q / \Delta t = (2.0 \times 10^{-3})(1.5 \times 10^4)(20) / (1.0)$

$\Delta Q / \Delta t = 6.0 \times 10^2$ cal /s

Example 2 One kilogram of ice is placed in an ice chest. The thickness of the sides of the chest is 4.0 cm and the total area of the sides of the ice chest is 0.50 m^2. If the temperature of the surroundings is 30°C, it takes 3h before all of the ice melts. Find the thermal conductivity for the material.

Solution

given: $\Delta T = 30°C$; $A = 0.50$ m^2 ; $t = 1.3 \times 10^5$ s ; L = 0.04 m

$Q = mL_f$

$Q = (1.0)(80) = 80$ kcal

$\Delta Q / \Delta t = kA \, \Delta T / L$

$80 / (1.3 \times 10^5) = k \, (0.50)(30) / 0.04$

$6.2 \times 10^{-4} = k \, (375)$

$k = 1.65 \times 10^{-6}$ kcal / m-s-°C

Example 3   The intensity of the radiation emitted by an object is 90 W.
             If the emissitivity of the material is 0.50 and the area is
             0.50 m^2, find the temperature of the radiating material.

Solution

given : $P = 90$ W ; $e = 0.50$; $A = 0.50$ m^2 ; $\sigma = 5.67 \times 10^{-8}$

$P = e\sigma AT^4$

$90 = (0.50)(5.67 \times 10^{-8})(0.50)\ T^4$

$6.3 \times 10^9 = T^4$

$T = 282$ K

Thermal Expansion of Materials

Example 1   A steel rod is 10 m long at a temperature of 10°C. Find
             the length of the rod at a temperature of 150°C.

Solution

given : $L_o = 10$ m ; $\Delta T = 140$°C ; $\alpha = 2.0 \times 10^{-5}$ 1/°C

$\Delta L = L_o\ \alpha\ \Delta T$

$\Delta L = (10)(2.0 \times 10^{-5})(140)$

$\Delta L = 2.8 \times 10^{-2}$ m

$L_{new} = 10.028$ m

Example 2   The density of mercury at 0°C is 13.6 g/cm^3. Find the
             density of mercury at a temperature of 90°C.

Solution

given : $\rho_{mercury}$ at 0°C $= 13.6$ g / cm^3

the mass remains constant as the volume increases

$\rho = m / V$

$\rho_1 / \rho_2 = V_2 / V_1$

$V_2 = V_1 + \Delta V$

$V_2 = V_1 + V_1 \beta\ \Delta T$

$\rho_1 / \rho_2 = V_1 + V_1 \beta\ \Delta T\ / (V_1)$

174

$$13.6 / \rho_2 = (1 + \beta \, \Delta T)$$

$$\rho_2 = 13.6 / [\, 1 + (0.18 \times 10^{-3})(90) \,]$$

$$\rho_2 = 13.6 / (1 + 1.62 \times 10^{-2}) = 13.4 \text{ g/cm}^3$$

Example 3 A 100 mL Pyrex beaker is filled to the brim with water at a temperature of 10°C. How much water should overflow if the two are heated to a temperature of 80°C ?

Solution

given : $V_o = 100$ mL ; $\beta_{water} = 0.21 \times 10^{-3}$ 1/°C ;

$\beta_{pyrex} = 0.9 \times 10^{-5}$ 1/°C

$$\Delta V_{water} = V_o \, \beta_{water} \, \Delta T$$
$$\Delta V_{water} = 100(0.21 \times 10^{-3})(70)$$
$$\Delta V_{water} = 1.47 \text{ mL}$$
$$\Delta V_{pyrex} = V_o \, \beta_{pyrex} \Delta T$$
$$\Delta V_{pyrex} = (100)(0.9 \times 10^{-5})(70) = 0.063 \text{ mL}$$
$$V_{overflows} = 1.41 \text{ mL}$$

Heat of Combustion

Example A person eats an eight ounce steak during the course of a meal. What is the energy content in kilocalories for the steak ?

Solution

given : m = 0.5 lb = 0.23 kg ;
heat of combustion = 1200 Kcal/kg

$$\Delta Q = mH$$
$$\Delta Q = (0.23)(1200) = 276 \text{ Kcal}$$

Solutions to Selected Problems from the Text

5. given : $h_g = 0.50$ cm ; $h_w = 4.0$ cm

    $\Delta Q / t = kA\,\Delta T / L$

    $\Delta Q / t_g = [(2.0 \times 10^{-3}) / 4.0]\, A\Delta T$

    $\Delta Q / t_g = (5.0 \times 10^{-4})\,(A\Delta T) =$

    $\Delta Q / t_w = [(2.5 \times 10^{-4}) / 0.50)]\, A\Delta T$

    $\Delta Q / t_w = (5.0 \times 10^{-4})\,(A\Delta T)$

    The two have the same ratio of heat conduction.

9. given : $h = 3.0 \times 10^{-1}$ cm ; copper - $k = 0.92$ cal/cm-s-°C ;
    $\Delta T = 50$ °C

    (a)　　　　$\Delta Q/\Delta t = kA\Delta T / h$

    　　　　　$\Delta Q/\Delta t = (0.92)(\pi)(10)(50) / (0.30)$

    　　　　　$\Delta Q/\Delta t = 4.8 \times 10^4$ cal /s or 48 kcal / s

    (b)　　　　$\Delta Q = mL_v$

    　　　　　$m = \Delta Q / L_v$

    　　　　　$m = (48$ kcal /s$)(300) / 540$ kcal / kg

    　　　　　$m = 27$ kg

14.　　　　$L_M = k\ (\text{R-val}) = (0.30)(38) = 11.4$ in

　　　　　$L_{SF} = k(\text{R-val}) = (0.30)(19) = 5.7$ in

19. given : emissitivity = 0.40 ; T = 373 K

    　　　$I = e\sigma T^4$

    　　　$I = (0.40)(5.67 \times 10^{-8})(373)^4$

    　　　$I = 4.4 \times 10^2$ W/m^2

24. given : $\Delta L = 0.021$ in ; $L_O = 1$ ft $= 12$ in

$\Delta L = L_O \alpha \Delta T$

$0.021 = 12 (0.66 \times 10^{-5})(\Delta T)$

$\Delta T = 265°F$

30. given : $\Delta T = 200°C$

$\Delta V = V_O 3\alpha \Delta T$

$\Delta V / V_O = 3\alpha \Delta T$

$\Delta V / V_O = 3 (2.0 \times 10^{-5})(200) = 0.012 = 1.2\%$

34. given : $\Delta T = 132°F$; $r = 2.5$ in ;

(a) $A = A_O (1 + 2\alpha \Delta T)$

$A = (\pi)(1.25)^2 [1 + (2)(1.3 \times 10^{-5})(132)]$

$A = 4.926$ in^2

(b) $A = L^2$

$4.926 = L^2$

$L = 2.219$ in.

39. given : 90% gas and 10% alcohol

$\Delta Q_T = (DVH)_g + (DVH)_{Al}$

$\Delta Q_T = (42 \text{ lb/ft}^3)(0.90 \text{ gal})(0.134 \text{ ft}^3/\text{gal}) (20.5 \times 10^3 \text{ Btu}) +$

$(49 \text{ lb/ft}^3)(0.10 \text{ gal})(0.134 \text{ ft}^3/\text{gal}) (11.5 \times 10^3 \text{ Btu})$

$\Delta Q_T = 11.14 \times 10^4$ Btu

the heat of combustion is slightly less than gas.

Review Questions

1. What are the distinctions between the three methods of heat transfer ?

    Conduction involves no net mass movement. Convection requires mass movement. Radiation requires no media.

2. How are thermal conductivity, thermal resistivity, and R-value related ?

    The thermal resistivity is the reciprocal of the thermal conductivity ($\rho = 1/k$).
    R-value is the product of the resistivity and the thickness of a material in inches. (R-value = $\rho L = L/k$.)

3. What does the R-value tell you ?

    Materials with equal R-values have equal insulating properties. Insulating requirements are rated in R-values and materials can be bought in terms of R-values. In this manner, the thickness of a particular material does not have to be specified.

4. If air is a good thermal insulator, why do we put insulation in the walls of a home ?

    With no insulation, the air between the walls would transfer heat by convection.

5. How does the intensity of radiation emitted by a body depend on temperature ?

    The intensity is proportional to the fourth power of the absolute temperature T^4. This means that if the absolute temperature is doubled, then the intensity is increased by a factor of 16.

6. On a molecular scale, why do materials generally expand with increase temperature ?

With increased temperature, the molecules of a material have more energy and greater vibrations about their equilibrium positions. The increased vibrations produces an expansion macroscopically.

7. What is the basic principle of cooking in a microwave oven.

Heat transfers by radiation. Water molecules have a resonance mode in the microwave range and absorb radiation energy. After that, the cooking or heat transfer is due to conduction or convection. The microwave penetrates a food only a few centimeters.

8. If the coefficient of area thermal expansion exactly equal to twice the coefficient of linear thermal expansion ?

NO. This is an approximation (first-order). The squared term is the equation is neglected.

9. Is the coefficient of volume thermal expansion for a liquid related to the linear coefficient of thermal expansion ?

NO. For a solid, to a good approximation, $\beta = 3\alpha$. The β's for liquids are generally larger and are not related to the α's for solids.

10. Does one get the complete heat of combustion for a fuel when it is burned ?

Only if the combustion is complete, which requires an oxygen atmosphere.

Sample Quiz

(Remove the quiz from the book and test your knowledge of the chapter material as through you were taking an in-class quiz. Check your answers with the key at the back of the Study Guide.)

Completion

1. The method of heat transfer that requires no material media is

   _____.

2. Two materials have the same R-value. The material with the greater thermal conductivity has the _____ thickness.

3. The intrinsic heat value per unit mass of a substance is called the

   _____.

Multiple Choice

___ 4. The warmth from an open fire is primarily due to
   A. convection   B. conduction
   C. radiation   D. none of the choices

___ 5. The coefficient of volume thermal expansion for a liquid is equivalent to
   A. α B. 2α C. 3α D. none of the choices

Problems

6. The thermal conductivity of a particular material is 0.25 Btu-in/ft^2-h-°F. What thickness of this material would give R-16 ?

7. A cube of material has a linear thermal coefficient of expansion of 3.0×10^{-5}°C^{-1}. What temperature change would cause a volume increase of 0.10 % ?

Chapter 18   Thermodynamics, Heat Engines and Heat Pumps

Sample Problems

The First Law of Thermodynamics and Thermodynamic Processes

Example 1   During an isothermal process, 500 kcal of heat are added to a system.
A. What is the change in internal energy ?
B. What is the amount of work done ?

Solution

given : isothermal process ; $\Delta Q = 500$ kcal

A.   Since the process is isothermal this means the change in temperature is zero. The change in internal energy is also zero.

B. $\Delta Q = \Delta U + W$
$\Delta Q = W$
500 kcal $= W$
$W = (500 \text{ kcal})(4.18 \times 10^3 \text{ J / kcal}) = 2.09 \times 10^6 \text{ J}$

Example 2   A gas is compressed from 2.0 m^3 to 1.2 m^3 at a constant pressure of 2.0 Atm while 10 kcal of heat are added.
A. Find the work done.
B. Find the change in internal energy.
C. Did the temperature increase, decrease, or remain the same ?

Solution

given : $\Delta V = -0.8 \text{ m}^3$; $P = 2$ Atm $= 2.026 \times 10^5 \text{ N/m}^2$;
$\Delta Q = 10$ kcal

A. $W = P\Delta V$

          $W = (2.026 \times 10^5)(-0.80)$

          $W = -1.62 \times 10^5$ J

B. $\Delta Q = \Delta U + W$

          4.18×10^4 J $= \Delta U + -1.62 \times 10^5$ J

          $\Delta U = 2.04 \times 10^5$ J

C.   Since the internal energy decreases, the temperature decreases.

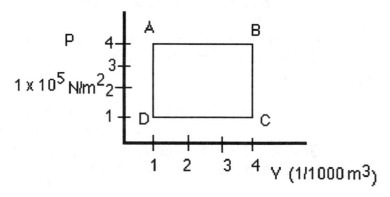

Example 3

A. Find the work done in the process ABCDA.
B. At what is the temperature the greatest ?
C. At what points are is the temperature the same ?
D. Calculate the ΔU from the entire process.
E. Calculate the change in internal energy for the process.

Solution

A. $W = P\Delta V$

          $W = (3.0 \times 10^5)(3.0 \times 10^{-3})$ (area of the rectangle)

          $W = 9.0 \times 10^2$ J

B.   B since the product of PV is the greatest.

C.   points A and C since the product of PV is the same.

D.   for the entire process the product of PV does not change - it starts at A and ends at A

E. $\Delta Q = \Delta U + W$

$\Delta Q = 9.0 \times 10^2$ J

$\Delta Q = (9.0 \times 10^2$ J$)(1$ J $/ 4.18$ cal$)$

$\Delta Q = 215$ cal

The Second and Third Laws of Thermodynamics

Example 1 Find the change in entropy when 10 g of steam at 100°C condenses to water at 100°C.

Solution

given : m = 10 g ; L_v = 540 cal/g ; T = 100°C = 373 K

$\Delta S = \Delta Q / T$

$\Delta S = mL / T$

$\Delta S = (10)(540) / 373$

$\Delta S = 14.5$ cal / K

Example 2 Find the change in entropy of a large body of water with the addition of 100 kcal of heat at a temperature of 20°C.

Solution

given : ΔQ = 100 kcal ; T = 20°C = 293 K

$\Delta S = \Delta Q / T$

$\Delta S = 100 / 293$

$\Delta S = 0.34$ kcal / K

Heat Engines

Example 1 The efficiency for a heat engine is 40%. If 200 kcal of heat is exhausted,
A. find the work done.
B. find the input heat.

<u>Solution</u>

given : eff = 0.40 ; Q_L = 200 kcal

A. $e = W_{output} / Q_H$
    $0.40 = W / (W + Q_L)$
    $0.40 = W / (W + 200)$
    $0.40\,W + 80 = W$
    $80 = W - (0.40)W$
    $80 = 0.60\,W$
    $W = 133$ kcal
    $W = (133$ kcal$) (4180\text{ J} / 1\text{ kcal}) = 5.57 \times 10^5$ J

B. $Q_H = Q_L + W$
    $Q_H = 200$ kcal + 133 kcal
    $Q_H = 333$ kcal

Example 2   Find the theoretical efficiency for a heat engine which operates between the temperatures of 100°C and 200°C.

<u>Solution</u>

given : T_C = 100°C = 373 K ; T_h = 200°C = 473 K

eff = 1 - (T_c / T_h)
eff = 1 - (373 / 473)
eff = 1 - 0.79
eff = 21 %

Heat Pumps and Refrigerators

Example   A 3.0 ton air conditioner is used for a home. How much heat is removed from the continuous use for one hour ?

<u>Solution</u>

given : 3.0 ton

(3.0 tons) (12,000 Btu / h) (1 h) = 36,000 Btu

Solutions from Selected Problems from the Text

5. given : $P = 1.013 \times 10^5 \text{ N/m}^2$; $\Delta V = -0.0150 \text{ m}^3$; $\Delta U = +500 \text{ J}$

first find the work done : $W = P\Delta V$
$$W = (1.013 \times 10^5 \text{ N/m}^2)(-0.015 \text{ m}^3)$$
$$W = -1.52 \times 10^3 \text{ J}$$

first law of thermodynamics : $\Delta Q = \Delta U + W$
$$\Delta Q = 500 \text{ J} + (-1.52 \times 10^3 \text{ J})$$
$$\Delta Q = -1.02 \times 10^3 \text{ J or } -0.24 \text{ kcal}$$

10. given : $U_1 = 147 \text{ J}$ at a temperature of 200 K ; $\Delta Q = 90 \text{ cal} = 377 \text{ J}$;
$$W = 20 \text{ J} = 84 \text{ J}$$
$$\Delta Q = \Delta U + W$$
$$377 \text{ J} = \Delta U + 84 \text{ J}$$
$$\Delta U = 293 \text{ J}$$
$$\Delta U = U_2 - U_1$$
$$293 \text{ J} = U_2 - 147 \text{ J}$$
$$U_2 = 440 \text{ J}$$

Since the temperature is directly proportional to the internal energy :

$$U_1 / U_2 = T_1 / T_2$$
$$147 / 440 = 200 / T_2$$
$$T_2 = 599 \text{ K}$$

15. given : $m_1 = 200 \text{ g}$ ice to water at 0°C ; $m_2 = 40 \text{ g}$ of water to steam at 100°C
$$\Delta Q_1 = mL_f$$
$$\Delta Q_1 = (200)(80) = 1.6 \times 10^4 \text{ cal or } 1.6 \text{ kcal}$$
$$\Delta S = \Delta Q / T$$

185

$$\Delta S_1 = (1.6 \times 10^4) / 273 \text{ K}$$
$$\Delta S_1 = 58.6 \text{ cal /K}$$
$$\Delta Q_2 = mL_v$$
$$\Delta Q_2 = - (40)(540) = -2.16 \times 10^4 \text{ cal} \quad (\text{- indicates heat must}$$
$$\text{be removed})$$
$$\Delta S_2 = (-2.16 \times 10^4) / 373$$
$$\Delta S_2 = -57.9 \text{ cal / K}$$

20. given : eff = 0.35 ; Q_h = 600 Btu ; Q_l = ? ; W = ?

(a)   eff = 1 - Q_l / Q_h

             0.35 = 1 - Q_l / 600

             Q_l / 600 = 0.65

             Q_l = 390 Btu

(b) Q_h = W + Q_l

             600 Btu = W + 390 Btu

             W = 210 Btu

25. given : eff = 0.25 ; P_{out} = 2.0 hp

             eff = P_{out} / P_{in}

             0.25 = 2.0 / P_{in}

             P_{in} = 8.0 hp

             (8.0 hp)[(550 ft-lb /s) / 1 hp] (3600 s / 1 h)

             1.6×10^7 ft-lb / h

30. given : $T_1 c$ = 373 K ; T_{1h} = 573 K ; T_{2c} = 100 K ; T_{2h} = 300 K

             $e_c = 1 - (T_c / T_h)$

             e_{c1} = 1 - (373 / 573) e_{c2} = 1 - (100 / 300)

             e_{c1} = 0.35 e_{c2} = 0.67

             e_{c1} = 35% e_{c2} = 67%

35. given : $Q_C = 150$ cal $= 628$ J ; W $= 150$ J

$cop = Q_C / W$

$cop = 628 / 150$

$cop = 4.19$

40. given : w $= 25$ ton ; convert water to ice at 0°C

$w = mg$

$w = [(\Delta Q / \Delta t) / L_f] \Delta t$

$w = (25$ ton$)(1.2 \times 10^4$ Btu/h / 1 ton$) (1$ h $) / 144$

$w = 2.1 \times 10^3$ lb $= 1.05$ ton

Review Questions

1. If heat is added to a system, what could possible happen ?

According to the first law of thermodynamics, the energy would go into doing work and/or the internal energy of the system.

2. Describe the following curves on a p-V plot for a perfect gas :
    A. isobar   B. isomet   C. isotherm

A. An isobar is a straight , horizontal line, constant pressure.
B. An isomet is a straight, vertical line, constant volume.
C. An isotherm is a curved line, or hyperbola since $p = NkT / V$
   which is of the form $y = a / x$.

3. Can the entropy of a system decrease ?

YES, but there will be a greater increase somewhere else in the universe such that the total entropy increases.

4. Suppose water spontaneously froze. How would this be described thermodynamically ?

As a violation of the second law. Water does not naturally or spontaneously freeze, since work most be done on the system to remove the heat or there is an entropy increase.

5. What is the "heat death of the universe" in terms of entropy ?

Thermal efficiency is the actual efficiency of a hear engine (the work out divided by the heat input). The Carnot efficiency is the ideal or maximum efficiency one could ever hope to achieve (but couldn't because the cycle is physically impossible in as much as no engine operating in a cycle could have perfect adiabats).

7. What is the difference between a two-stroke cycle and a four-stroke cycle in terms of piston motion of an engine ?

A two-stroke cycle has one up and down piston motion per cycle. A four-stroke cycle has two up and down motions per cycle.

8. Define a heat pump in thermodynamic terms.

A heat pump is a device that transfers heat from a low-temperature reservoir to a high temperature reservoir. This requires work input.

9. What is a throttling process in refrigeration ?

In the refrigeration cycle, a high-pressure liquid expands through a small opening or expansion valve and partial vaporization occurs. This lowers the temperature at the expense of internal energy.

10. Where does a heat pump used to heat a house get its heat ?

Heat is exchanged or taken from the outside air or a water reservoir. In effect, heat is pumped from a low-temperature reservoir to a high temperature reservoir, which requires work input (usually electrical energy).

Sample Quiz

(Remove the quiz from the book and test your knowledge of the chapter materials as though you were taking an in-class quiz. Check your answers with the key at the back of the Study Guide.)

Completion

1. No work is done on or by a gas is a _____ process.

2. In a Carnot process, the efficiency from a low-temperature reservoir to a high- temperature reservoir would not violate the _____ law of thermodynamics.

Multiple Choice.

____4. A change in the temperature of a gas system can occur in an
    A. isometric process   C. adiabatic process
    B. isobaric process   D. all of the choices

____5. The coefficient of performance
    A. is always less than one
    B. can be greater than one
    C. is the reciprocal of the efficiency
    D. none of the above

Problems

6. Twenty grams of water at 100 °C are converted to steam. What is the change in entropy ?   Is it an increase or decrease ?

7. A heat engine absorbs 100 J of energy from a 400°C reservoir and rejects 58 J to an 80°C reservoir. What is
   A. the thermal efficiency ?
   B. the Carnot efficiency ?

Chapter 19   Electrostatics

Sample Problems

Electric Charge and Force

Example 1   The net charge on a substance is 6.0 μC. How many
more protons are there than electrons ?

Solution

given : $q = 6.0 \times 10^{-6}$ C ; charge of proton $= 1.6 \times 10^{-19}$ C

$n = (6.0 \times 10^{-6}$ C) $/ (1.6 \times 10^{-19}$ C $/$ proton)
$n = 3.75 \times 10^{13}$ protons than electrons

Example 2   Two 6.0 μC charges are placed 50 cm apart. Find the
electrostatic force between the charges.

Solution

given : $q_1 = q_2 = 6.0 \times 10^{-6}$ C ; $r = 0.50$ m ;
$k = 9.0 \times 10^9$ N-m^2/C^2

$F = kq_1\, q_2\, / r^2$
$F = (9.0 \times 10^9)\, (6.0 \times 10^{-6})(6.0 \times 10^{-6})\, / (0.50)^2$
$F = 1.3$ N

Example 3   Three 6.0 μC are located at the vertices of an equilateral
triangle 50 cm on a side.   Find the net force on Q_1.

Solution

given : $k = 9.0 \times 10^9$ N-m^2 / C^2 ; $q = 6.0 \times 10^{-6}$ C ;
$r = 0.50$ m
From example 2, the force between each of the two
charges is

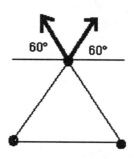

Since the two forces are equal, by symmetry, the forces in the x component cancel. Therefore the net force is in the y- component.

F = 2F sin60

F = 2 (1.3) sin60

F = 2.3 N

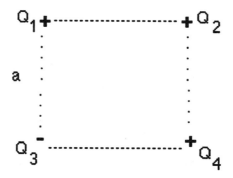

Example 4   Four 6.0 μC charges are located at the vertices of an equilateral triangle 50 cm on a side. Find the net force in Q_1.

<u>Solution</u>

given : $k = 9.0 \times 10^9$ N-m^2/C^2 ; $q = 6.0 \times 10^{-6}$ C ; r = 0.50 m

From example 2, the magnitude of the forces F_{12} and F_{13} are the same since the charges are the same and the distance between the charges is the same. F_{14} is different since the distance between the charges is different.

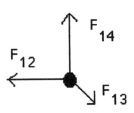

$F_{14} = (9.0 \times 10^9) (6.0 \times 10^{-6})(6.0 \times 10^{-6}) / (0.71)^2 = 0.65$ N

$F_x = F_{12} - F_{13} \cos 45$

$F_x = 1.3 - (0.65)\cos 45 = 0.84$ N

$F_y = F_{14} - F_{13} \sin 45$

$F_y = 1.3 - (0.65)\sin 45 = 0.84$ N

$F_r^2 = F_x^2 + F_y^2$

$F_r^2 = 0.84^2 + 0.84^2$

$F_r = 1.19$ N   45° above the **-x** axis

Electric Field

Example 1   A 5.0 C charge is located on the x-axis at the point (2,0) m.
Find the electric field at the origin.

Solution

given : $r = 2.0$ m ; $q = 5.0 \times 10^{-6}$ C ; $k = 9.0 \times 10^9$ N-m^2/C^2

$E = kq / r^2$

$E = (9.0 \times 10^9)(5.0 \times 10^{-6}) / 2.0^2$

$E = 1.1 \times 10^4$ in the **+x** direction.

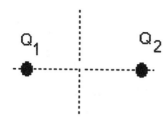

Q_1 Q_2

Example 2   Two 5.0 μC point charges shown above are located on
              the x-axis at x = ±2 m.
              A. Find the electric field at the origin if the charges are
                 both positive or negative.
              B. Find the electric field at the origin if Q_1 is (+) and
                 Q_2 is (-).

<u>Solution</u>

given : $q = 5.0 \times 10^{-6}$ C ; $r = 2.0$ m ; $k = 9.0 \times 10^9$ N-m^2/C^2

A.   Since the vectors for the electric field are in opposite
         directions the net electric field is zero.

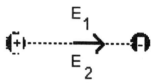

B. $E = E_1 + E_2$ (since the vectors are in the same direction)

         $E_1 = 2 \, (9.0 \times 10^9) \, (5.0 \times 10^{-6}) / (2.0)^2$

         $E_1 = 2.25 \times 10^4$ N/C in the **+x** direction

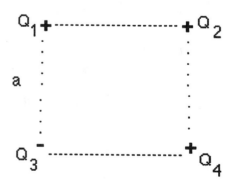

Example 3   Four point charges are located at the vertices of a square
              as shown above. If a = 50 cm and the magnitude of
              each charge is 50 μC, find the electric field at the center of
              the square.

$$\text{given}: \ k = 9.0 \times 10^9 \ ; \ q = 50 \times 10^{-6} \ C$$

first find r ; r = one half the diagonal of the square

$$d^2 = a^2 + a^2$$
$$d^2 = (0.50)^2 + (0.50)^2$$
$$d = 0.7 \ m$$
$$r = 0.35 \ m$$

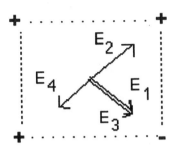

By inspection the electric fields from charges 2 and 4 cancel.

$$E = E_1 + E_3$$
$$E = 2 \, E_1$$
$$E = 2 \, (9.0 \times 10^9)(50 \times 10^{-6}) \, / \, (0.35)^2$$
$$E = 7.3 \times 10^6 \ N/C \quad \text{toward the - charge.}$$

Electric Potential Energy and Electric Potential

Example 1   Two point electrons at rest are located 50 cm apart.
A. Find the potential energy of the system.
B. If the charges each move 5.0 cm (they are 60 cm apart), find the potential energy of the system.
C. What happens to the potential energy ?

Solution

$$\text{given}: q = -1.6 \times 10^{-19} \ C \ ; \ r = 0.50 \ m \ ;$$
$$k = 9.0 \times 10^9 \ N\text{-}m^2 / C^2$$

A. $PE = kq_1q_2/r$

$PE = (9.0 \times 10^9)\ (-1.6 \times 10^{-19})^2 / (0.50)$

$PE = 4.6 \times 10^{-28}\ J$

B. $PE = (9.0 \times 10^9)(-1.6 \times 10^{-19})^2 / (0.60)$

$PE = 3.8 \times 10^{-28}\ J$

C. The potential energy decrease, therefore the K of the particles increases.

$PE_1 + K_1 = PE_2 + K_2$

$4.6 \times 10^{-28}\ J\ + 0 = 3.8 \times 10^{-28}\ J + K_2$

$K_2 = 1.2 \times 10^{-28}\ J$ between the two electrons

Example 2   Two $2.0\ \mu C$ point charges are located on the x-axis at $x = \pm\ 3.0$ cm.
A. Find the electric potential at the origin.
B. Find the electric potential at the point (0 , 4) cm.

Solution

given : $k = 9.0 \times 10^9$; $q_1 = q_2 = 2.0 \times 10^{-6}$; $r = 0.03$ m

A. $V = V_1 + V_2$

$V = (kq_1 / r) + (kq_2 / r)$

$V = (9.0 \times 10^9)(2.0 \times 10^{-6}) / (0.03)\ +$
$\qquad (9.0 \times 10^9)(2.0 \times 10^{-6}) / (0.03)$

$V = 6.0 \times 10^5\ V\ + 6.0 \times 10^5\ V$

$V = 1.2 \times 10^6\ V$

B. $V = V_1 + V_2$

$V = (9.0 \times 10^9)(2.0 \times 10^{-6}) / 0.05\ +$
$\qquad (9.0 \times 10^9)(2.0 \times 10^{-6}) / (0.05)$

$V = 3.6 \times 10^5\ V + \ 3.6 \times 10^5\ V$

$V = 7.2 \times 10^5\ V$

Solutions from Selected Problems from the Text

5. given : $r = 0.20$ m ; $q_1 = -1.6 \times 10^{-19}$ C ; $q_2 = +0.80$ C

    A. $F = kq_1 q_2 / r^2$

            $F = (9.0 \times 10^9)(-1.6 \times 10^{-19})(0.80) / (0.20)^2$

            $F = -2.9 \times 10^{-8}$ N or 2.9×10^{-8} N toward the + charge

    B. $F = ma$

            $(2.9 \times 10^{-8}) = (9.1 \times 10^{-31})$ a

            $a = 3.2 \times 10^{22}$ m/s^2 toward the + charge

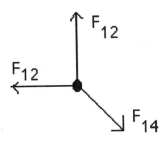

10.

    (a) $F_{12} = kq_1 q_2 / r^2$

            $F_{12} = (9.0 \times 10^9)(2.0 \times 10^{-6})(3.0 \times 10^{-6}) / (0.30)^2 = 0.60$ N

            $F_{13} = 0.60$ N (since the product of the charges is the
            same and r is the same)

            $F_{14} = (9.0 \times 10^9)(2.0 \times 10^{-6})(2.0 \times 10^{-6}) / (0.42)^2 = 0.20$ N

            $F_x = 0.60 - (0.20)(\cos 45°) = 0.46$ N - x direction

            $F_y = 0.46$ N + y direction

            $F_r^2 = F_x^2 + F_y^2$

            $F_r = 0.65$ N $45°$ above the -x axis

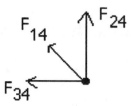

(b)

$$F_x = F_{34} + F_{14}\cos 45$$
$$F_x = 0.60 + (0.20)\cos 45 = -0.74\ N$$
$$F_y = 0.60 + (0.20)\sin 45 = 0.74\ N$$
$$F_r = (F_x^2 + F_y^2)^{1/2}$$
$$F_r = (0.74^2 + 0.74^2)^{1/2} = 1.05\ N$$
$$\text{Tan}\varnothing = 0.74 / 0.74$$
$$\varnothing = 45°\ \text{above the -x axis}$$

15. given : Q - charge for the sphere is 0.25 μC ; area = 0.80 m

    (a) $E = \sigma / \varepsilon = (Q / A) / \varepsilon$

           $E = [(0.25 \times 10^{-6}) / (0.80)] / (8.85 \times 10^{-12})$

           $E = 3.5 \times 10^5\ N/C$ away from the surface

    (b) $A = \pi r^2$

           $0.80 = \pi r^2$

           $r = 0.50\ m$

           Any location inside of the sphere the electric field is equal to zero.

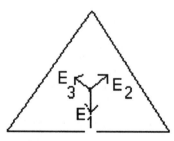

20.

    given : $q_1 = q_2 = q_3 = -4.0$ μC ; each side has a measure of 0.40 m

        $E_x = E_2\cos 30 - E_3\cos 30$

        $E_x = 0$ since q / r^2 is the same for each vector.

$$E_y = 2kQ\cos 60 / r^2 - kQ / r^2$$
$$E_y = 0 \qquad ; \text{ since } 2(\cos 60) = 1$$

Therefore the electric field at the center is zero.

25. given : $q_1 = q_2 = q_3 = -4.0 \, \mu C$; each side has a measure of 0.40 m

$$\Delta U = U_{12} + U_{23} + U_{13}$$
$$\Delta U = 3 \, (U_{12}) \text{ since } U_{12} = U_{23} = U_{13}$$
$$\Delta U = 3 \, kq_1q_2 / r$$
$$\Delta U = [3 \, (9.0 \times 10^9)(-4.0 \times 10^{-6})^2 / 0.40$$
$$\Delta U = 1.9 \, J$$

30. given : $V = 1.0 \times 10^3 \, V$; $q = e = -1.6 \times 10^{-19} C$
$$\Delta V = K / q$$
$$K = \Delta V \, q$$
$$K = (1.0 \times 10^3)(1.6 \times 10^{-19}) = 1.6 \times 10^{-16} \, J$$

Review Questions

1. What is electric charge ?

The fundamental property associated with subatomic particles, the electron and proton. There are two types of charges, positive and negative.

2. What do good conductors and thermal conductors have in common ?

These are generally metals and have electron that are "free" to move around. The electrons are not permanently bound to a particular molecule or atom.

3. Why does a balloon stick to a wall after being rubbed on a sweater or hair ?

   The balloon is charged by the rubbing contact or friction, and the charged balloon induces an opposite charge on the wall, which provides an attractive electrical force.

4. What is an electric field ?

   The force per unit charge, $E = F/q$. The electric field tells what force would be experienced by a charge placed at a particular location.

5. How does the electrical force compare to the gravitational force ?

   The electrical force is on the order of 10^{40} times stronger than the gravitational force.

6. If a positively charged rod is brought near a negatively charged electroscope, what happens and why ?

   The leaves of the electroscope collapse as the negative charge in the diverged leaves is attracted toward the positively charged rod near the bulb.

7. What gives rise to electric potential energy ?

   Work done against the electric force.

8. Distinguish between electric potential energy and electric potential.

   Electric potential is the energy per unit charge.

9. What are two equivalent units for electric potential ?

    J / C or volt (V)

10. What is an electron volt ?

    A unit of energy defined by the amount of energy gained by an electron when accelerated through a potentials difference of one volt.

Sample Quiz

(Remove the quiz from the book and test your knowledge of the chapter materials as though you were taking an in-class quiz. Check your answers with the key at the back of the Study Guide.)

Completion

1. A positive charge between a negative and positive charge would experience a force _____ the _____ charge.

2. The force per unit charge is _____.

3. The electric potential per unit charge is commonly measured in _____.

Multiple Choice

____4. A large conductor that can receive or supply electrons without affecting its own electrical condition is called
   A. a semiconductor   C. electrical ground
   B. polarization   D. a Coulomb reservoir

____5. If two negative charges are moved closer together
   A. the electric field decreases
   B. the electric potential increases
   C. there is no change in the voltage difference
   D. none of the above

Problems

6. Two charges , one positive and one negative with a magnitude of 60 μC are located positions (-3 , 0) m and (3 , 0) m, respectively.
   A. What is the electric field at the origin ?
   B. What is the electric potential at the origin ?

7. An electron is accelerated through a potential difference of 100 V. How much energy does the electron require ? (Express your answer in standard units.)

Chapter 20

Capacitance and Dielectrics

Sample Problems

Capacitance and Capacitors

Example 1 A parallel plate capacitor consists of two plates whose area is $0.25 \ m^2$ separated by 10 cm of air.
A. Find the capacitance of the capacitor.
B. If the capacitor is connected to a 10 V source, find the charge stored on the capacitor.

Solution

given : $A = 0.25 \ m^2$; $d = 0.10 \ m$; $V = 10 \ V$

A. $C = \varepsilon A / d$
$C = (8.85 \times 10^{-12}) \ (0.25) / 0.10$
$C = 2.2 \times 10^{-11} \ F$

B. $C = Q / V$
$Q = CV$
$Q = (2.2 \times 10^{-11})(10)$
$Q = 2.2 \times 10^{-10} \ C$

Example 2 A parallel plate capacitor has 10 μC of charge when connected to a 20 V source. If the separation between the plates is 10 mm, find the
A. capacitance.
B. area of the plates.

Solution

given : $Q = 10 \times 10^{-6} \ C$; $V = 20 \ V$; $d = 10 \times 10^{-3} \ m$

A. $C = Q / V$
$C = (1 \times 10^{-6}) / 20$
$C = 5 \times 10^{-8} \ F$

B. $C = \varepsilon A / d$

$$5 \times 10^{-8} = (8.85 \times 10^{-12}) \, A / (10 \times 10^{-3})$$
$$(5.0 \times 10^{-8}) / (8.85 \times 10^{-10}) = A$$
$$A = 56 \, m^2$$

Energy of a Charged Capacitor

Example 1   A 10 μF capacitor has a charge of 100 μC stored on its plates.
A. Find the potential difference between the plates.
B. Find the energy stored by the capacitors.

Solution

given : $C = 10 \, \mu F$; $q = 100 \, \mu C$

A. $C = Q / V$
$V = Q / C$
$V = (100 \, \mu C / 10 \, \mu F)$
$V = 10 \, V$

B. $U = (1/2)CV^2$
$U = (1/2)(100 \times 10^{-6})(10)^2$
$U = 5.0 \times 10^{-3} \, J$

Example 2   A capacitor stores 100 mJ of energy when connected to a 10 V battery.
A.   Find the capacitance.
B.   Find the charge stored by the capacitor.

Solution

given : $U = 100 \times 10^{-3} \, J$; $V = 10 \, V$

A. $U = (1/2)CV^2$
$100 \times 10^{-3} = (1/2) \, C \, (10)^2$
$100 \times 10^{-3} = 50 \, C$
$2 \times 10^{-3} = C$

B. $C = Q / V$
$(2 \times 10^{-3}) = Q / 10$
$Q = 2.0 \times 10^{-2} \, C$

Dielectrics and Dielectric Constants

Example   A parallel plate capacitor is charged to a 10 V battery.
The plates are separated 3.0 cm by teflon. The area of the
plates is 30 cm^2.
A. Find the capacitance.
B. Find the charge on the capacitor.
C. Find the energy stored on the capacitor.

Solution

given : V = 10 V ; k for teflon = 2.1 ;
A = 30 cm^2 = 0.03 m^2 ; d = 0.03 m

A. $C = k \, \varepsilon \, A / d$
$C = (2.1)(8.85 \times 10^{-12}) \, (0.03) / (0.03)$
$C = 1.86 \times 10^{-11}$ F

B. $C = Q / V$
$CV = Q$
$Q = (1.86 \times 10^{-11})(10)$
$Q = 1.86 \times 10^{-10}$ C

Capacitors in Series and Parallel

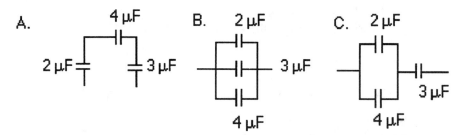

Example 1   Find the total capacitance in each of the following.

Solution

A. $1 / C_t = 1 / C_1 + 1 / C_2 + 1 / C_3$
$1 / C_t = 1/2 + 1/4 + 1/3$
$C_t = 0.92 \, \mu F$

B.　　$C_t = C_1 + C_2 + C_3$
　　　　$C_t = 2 + 4 + 3$
　　　　$C = 9.0 \ \mu F$

C.　　first find the equivalent capacitance for the parallel
　　　combination
　　　$C_{eq} = C_1 + C_2$
　　　$C_{eq} = 2 + 4 = 6 \ \mu F$

　　　$1/C_t = 1/C_3 + 1/C_{eq}$
　　　$1/C_t = 1/3 + 1/6$
　　　$C_t = 2 \ \mu F$

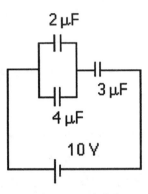

2 μF

3 μF

4 μF

10 V

Example 2　Find the charge and potential for each capacitor when
　　　　　charged by the 10 V battery.

<u>Solution</u>

First find the equivalent capacitance for the combination.
Note the capacitance is the same as in example 1.

$C_t = Q_t / V_t$
$Q_t = (2 \ \mu F)(10 \ V) = 20 \ \mu C$
$Q_{3\mu F} = Q_{2\mu F}$ and $4 \ \mu F = 20 \ \mu C$
$V_3 = Q_3 / C_3$
$V_3 = 20 \ \mu C / 3 \ \mu F = 6.7 \ V$
$V_{2 \ and \ 4} = 20 \ \mu C / 6 \ \mu F = 3.3 \ V$
$Q_2 = C_2 V_2$
$Q_2 = (2 \ \mu F)(3.3 \ V) = 6.6 \ \mu C$
$Q_4 = (4.0 \ \mu F)(3.3 \ V) = 13.2 \ \mu C$

Capacitors Charging and Discharging

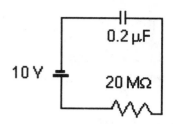

0.2 μF

10 V

20 MΩ

Example 1   A 20 MΩ resistor is connected in series to a 0.20 μF capacitor. A 10 V source is applied to the combination.
A. Find the time constant.
B. Find the initial voltage across the resistor and capacitor.
C. Find the final voltage across the resistor and capacitor.
D. Find the total charge stored on the capacitor.
E. Find the charge and the current in the circuit after 2.0 s.

Solution

given : $C = 2.0 \times 10^{-7}$ F ; $R = 20 \times 10^{6}$ Ω ; $V = 10$ V

A. $\tau = RC$
   $\tau = (2.0 \times 10^{7})(2.0 \times 10^{-7}) = 4.0$ s

B. $V_c = 0$ since it uncharged ; $V_r = 10$ V since $V_r + V_c = 10$

C. $V_c = 10$ V and $V_r = 0$ since the current ceases flow in the circuit.

D. $U = (1/2)CV^2$
   $U = (1/2)(2.0 \times 10^{-7})(10)^2 = 1.0 \times 10^{-5}$ J

E. $Q = Q_{max} (1 - e^{T/\tau})$
   $Q_{max} = CV$
   $Q_{max} = (2.0 \times 10^{-7})(10) = 2.0 \times 10^{-6}$ C
   $Q = 2\mu C (1 - e^{-2/4})$
   $Q = 2\mu C (1 - e^{-0.5})$
   $Q = 0.78 \,\mu C$

   $i = i_{max} (e^{T/\tau})$
   $i_{max} = (10) / (2.0 \times 10^{7}) = 5.0 \times 10^{-7}$ A
   $i = (5.0 \times 10^{-7}) (2.718^{-0.5})$
   $i = 3.0 \times 10^{-7}$ A

Example 2　A 20 MΩ resistor is connected to a 0.20 μF capacitor
which has 2.0 mJ of energy.
A. Find the time constant.
B. Find the initial voltage across the resistor and capacitor.
C. Find the final voltage across the resistor and capacitor.
D. Find the total charge stored on the capacitor.
E. Find the charge and the current in the circuit after 2.0 s.

Solution

A.　$\tau = RC = (20 \times 10^6)(0.20 \times 10^{-6}) = 4.0\ s$

B.　$U = (1/2)CV^2$
$2.0 \times 10^{-3} = (1/2)(2.0 \times 10^{-7})(V^2)$
$V = 140\ V$
V_c and $V_r = 1.4 \times 10^2\ V$

C.　After a long time the capacitor becomes discharged and
the voltage for both the resistor and capacitor becomes
zero.

D.　$C = Q/V$
$Q = CV$
$Q = (2.0 \times 10^{-7})(1.4 \times 10^2)$
$Q = 2.8 \times 10^{-5}\ C$

E.　$Q = Q_{max}\ (e^{T/\tau})$
$Q = (2.8 \times 10^{-5})\ e^{-0.5}$
$Q = 1.7 \times 10^{-5}\ C$

$i = i_{max}\ (e^{T/\tau})$
$i = (1.4 \times 10^2 / 2.0 \times 10^7)\ (e^{-0.5})$
$i = (7.0 \times 10^{-6})\ (0.61)$
$i = 4.3 \times 10^{-6}\ A$

Solutions from Selected Problems from the Text

5. given : $d = 1.0 \times 10^{-3}$ m ; $A = lw = (1.0 \text{ m})(0.75 \text{ m}) = 0.75$ m^2

$C = \varepsilon A / d$
$C = (8.85 \times 10^{-12})(0.75) / (1.0 \times 10^{-3})$
$C = 6.6 \times 10^{-9}$ F

10. given : $C = 20$ pF $= 20 \times 10^{-12}$ F ; $V = 12$ V

$U = (1/2)CV^2$
$U = (1/2)(20 \times 10^{-12})(12)^2$
$U = 2.8 \times 10^{-9}$ J

15. given : $d = 2.0$ mm $= 2.0 \times 10^{-3}$ m ; $V = 120$ V
$\mu = (1/2)\varepsilon E^2$
$\mu = (1/2)(\varepsilon)(V/d)^2$
$\mu = (1/2)(8.85 \times 10^{-12})[120 / (2.0 \times 10^{-3})]^2$
$\mu = 1.6 \times 10^{-2}$ J/m^3

20. given : $K = 4.0$; $A = 0.50$ m^2 ; $d = 1.4 \times 10^{-3}$ m ; $V = 12$ V
    A. $C = K \varepsilon A / d$
           $C = (4.0)(8.85 \times 10^{-12})(0.50) / 1.4 \times 10^{-3}$
           $C = 1.3 \times 10^{-8}$ F

    B. $U = (1/2)CV^2$
           $U = (1/2)(1.3 \times 10^{-8})(12)^2$
           $U = 9.4 \times 10^{-7}$ J

25. given : $C_1 = 0.50\ \mu F$; $C_2 = 1.0\ \mu F$; $C_3 = 1.5\ \mu F$

To obtain the minimum capacitance, the capacitors would be connected in series.

$$1/C_t = 1/C_1 + 1/C_2 + 1/C_3$$
$$1/C_t = 1/0.5 + 1/1.0 + 1/1.5$$
$$1/C_t = 2 + 1 + 2/3$$
$$1/C_t = 3.67$$
$$C_t = 0.27\ \mu F$$

To obtain the maximum capacitance, the capacitors would be connected in parallel.

$$C_t = C_1 + C_2 + C_3$$
$$C_t = 0.5\ \mu F + 1.0\ \mu F + 1.5\ \mu F$$
$$C_t = 3.0\ \mu F$$

30. given : $C_1 = 2.0\ \mu F$; $C_2 = ?$; $C_3 = 3.0\ \mu F$; $C_t = 4.0\ \mu C$

$$C_t = C_3 + C_{eq}$$
$$4.0\ \mu F = 3.0\ \mu F + C_{eq}$$
$$C_{eq} = 1.0\ \mu F$$

$$1/C_{eq} = 1/C_1 + 1/C_2$$
$$1/1.0\ \mu F = 1/2.0\ \mu F + 1/C_2$$
$$1 - 1/2 = 1/C_2$$
$$1/2 = 1/C_2$$
$$C_2 = 2.0\ \mu F$$

35. given : $t = 3\tau$
Capacitor Charging

$$V = V_{max}\ (1 - e^{-t}/RC)$$
$$V = V_{max}\ (1 - e^{-3RC}/RC)$$
$$V = V_{max}\ (1 - 0.05)$$
$$V = (0.95)V_{max}$$

Capacitor Discharging

$$V = V_{max}(e^{-t/RC})$$
$$V = V_{max}(e^{-3}RC/RC)$$
$$V = V_{max}(0.05)$$
$$V = (0.05)V_{max}$$

Review Questions

1. What is capacitance ?

    The ratio of the charge transferred to applied voltage $(C = Q/V)$.

2. The capacitance of a parallel-plate capacitor depends on what two geometric considerations ?

    The area of the plates and the separation distance of the plates $(C = \varepsilon_0 A/d)$.

3. What is the permittivity of free space ?

    A fundamental constant with a value of $8.85 \times 10^{-12} C^2/N\text{-}m^2$.

4. How does the energy stored in a capacitor vary with voltage ?

    The energy is proportional to the square of the voltage, $U = (1/2)CV^2$, so when the voltage is doubled, the energy increases by a factor of four.

5. What is a dielectric and what is dielectric strength ?

    A dielectric is an insulator. The voltage at which electrical breakdown occurs is the dielectric strength of the material.

6. What is the dielectric in electrolytic capacitors ?

   A thin oxide film that formed electrolytically on the surface of one plate, usually aluminum foil. The oxide films are on the order of a few millionths of an inch thick.

7. To have the same charge on several capacitors in a battery circuit, how should they be connected ?

   In series.

8. When several capacitors are connected in parallel in a battery circuit, what is the voltage drop across each capacitor ?

   The same for each capacitor and equal to the voltage of the battery.

9. Given several capacitors, how could you connect them to get the minimal capacitance ?

   In series. When connected in series, the equivalent capacitance is less than that of the smallest capacitor.

10. By increasing the time constant in an RC circuit, how are the charging and discharging rates of the capacitor affected ?

    The capacitor would charge up more slowly and discharge more slowly. It would take a longer time (one time constant) for the capacitor to charge to 63% of its maximum value, and a longer time (one time constant) to discharge to 37% of its maximum value.

Sample Quiz

(Remove the quiz from the book and test your knowledge of the chapter materials as though you were taking an in-class quiz. Check your answers with the key at the back of the Study Guide.

Completion

1. If the separation distance of the plates of a parallel plate capacitor is increased, the capacitance is _____.

2. The dielectric constant of a material is the ratio of the capacitances of a capacitor _____ a dielectric.

3. The greatest capacitance can be obtained by connecting several capacitors in _____.

Multiple Choice

___4. Energy is stored in a capacitor with a dielectric as a result of work done
    A. on electric dipoles   C. by the leaking current
    B. on the permittivity   D. none of the choices

___5. The charging rate of a capacitor depends on
    A. the dielectric strength   C. the value of the capacitance
    B. the permittivity of free space   D. the ratio of R / C

Problems

6. A parallel-plate capacitor with a plate of area 0.50 m^2 and a
   separation distance of 0.10 mm contains a dielectric with a dielectric
   constant of 2.5. If the capacitor is connected to a 6.0 V battery,
       A. how much energy is stored in the capacitor ?
       B. how much charge is on the plates of the capacitors ?
       C. what is the capacitance ?

7. Two capacitors with values of 2 μF and 4.0 μF are connected in
   series with each other. The combinations of resistors is placed in
   series with a 500 kΩ resistor. If the arrangement is connected to a
   12 V battery,
       A. find the equivalent capacitance.
       B. the time needed for the capacitors to charge to two-thirds of
          their maximum value.

Chapter 21   Current, Resistance, and Power

Sample Problems

Electric Current

Example   A 5.0 A current passes through a wire for a time of 10 min.
                A. How much charge passes through the wire during this
                     time ?
                B. How many electrons does this represent ?

Solution

given : $i = 5.0$ A ; $t = 10$ min $= 600$ s

A. $q = it$
          $q = (5.0)(600)$
          $q = 3.0 \times 10^3$ C

B. 3.0×10^3 C $/ (1.6 \times 10^{-19}$ C $/$ electron$)$
          $n = 1.9 \times 10^{22}$ electrons

Ohm's Law

Example 1   A 10 V source is connected to a 20 Ω resistor. Calculate
                the current in the resistor.

Solution

given : $V = 10$ V ; $R = 20$ Ω

$V = iR$
$10 = i\,(20)$
$i = 0.50$ A

Example 2   A light bulb has a resistance of 360 Ω. Calculate the
                current in the bulb if the voltage in the bulb is
                A. 110 V
                B. 100 V

Solution

$$\text{given}: \ R = 360 \ \Omega \ ; V_1 = 110 \ V \ ; \ V_2 = 100 \ V$$

A. $V = iR$
$110 = i \ (360)$
$i = 0.31 \ A$

B. $100 = i \ (360)$
$i = 0.28 \ A$

Resistance and Resistivity

Example 1   A copper wire has a length of 20 m and a radius of 1.0 mm. Find the resistance of the wire.

Solution

$$\text{given}: \ \rho = 1.7 \times 10^{-8} \ \Omega\text{-m} \ ; L = 20 \ m \ ; r = 1.0 \times 10^{-3} \ m$$

$R = \rho L / A$
$R = (1.7 \times 10^{-8})(20) / (\pi)(1.0 \times 10^{-3})^2$
$R = 0.11 \ \Omega$

Example 2   An aluminum wire has the same resistance as a copper wire of equal length. Calculate the radius of the wire.

Solution

$$\text{given}: \ \rho_{copper} = 1.7 \times 10^{-8} \ \Omega\text{-m} \ ; \rho_{Al} = 2.8 \times 10^{-8} \ \Omega\text{-m}$$

$R = \rho L / A$
$R_c = R_{al}$
$\rho L_c / A_c = \rho L_{Al} / A_{Al}$
$(1.7 \times 10^{-8}) / \pi \, r_{al}^2 = (2.8 \times 10^{-8}) / \pi \, r_c^2$
$(1.3 \times 10^{-4}) / (1.7 \times 10^{-4}) = r_{al} / r_c$
$0.76 \, r_{copper} = r_{aluminum}$

Example 3   A nichrome filament has a resistance of 10 Ω at room
             temperature.   What temperature is needed for the
             resistance to increase by 10% ?

Solution

given : $\alpha = 0.44 \times 10^{-3}$ Ω-m ; R = 1.1 R_o

$R = R_o (1 + \alpha\Delta T)$
$1.1\ R_o = R_o (1 + \alpha\Delta T)$
$1.1 = 1 + \alpha\Delta T$
$0.1 = \alpha\Delta T$
$0.1 = (0.44 \times 10^{-3})\Delta T$
$\Delta T = 227\ °C$
$T = 247°C$

Electric Power

Example 1   A 30 Ω resistor is connected to a 120 V source.
             A. Calculate the current in the resistor.
             B. Calculate the power in the resistor.
             C. How much energy would be expended in 1 h ?

Solution

given : R = 30 Ω ; V = 120 V ; t = 3600 s

A. $V = iR$
      $120 = i\ (30)$
      $i = 4.0\ A$

B. $P = V^2 / R$
      $P = 120^2 / 30$
      $P = 480\ W$

C. $P = E / t$
      $480 = E / 3600$
      $E = 1.7 \times 10^6\ J$

218

Example 2   A 1500 W hair dryer is connected to a 120 V source.
            A. Calculate the resistance of the coil.
            B. How much heat is produced in 10 min ?
Solution

given : $P = 1500$ W ; $V = 120$ V ; $t = 10$ min $= 600$ s

A. $P = V^2 / R$
      $1500 = 120^2 / R$
      $R = (1.44 \times 10^4) / 1500$
      $R = 9.6 \, \Omega$

B. $P = E / t$
      $1500 = E / 600$ s
      $E = 9.0 \times 10^5$ J

Example 3   An electric heater is rated at 5000 W at 220 V. How much
            power would be expended if the voltage is reduced by
            15% ?

Solution

given : $V_1 = 220$ V ; $P_1 = 5000$ W ; $V_2 = 0.85 \, V_1 = 187$ V ;
        $P_2 = ?$

$P = V^2 / R$
the resistance remains the same

$P_1 / P_2 = (V_1 / V_2)^2$
$5000 / P_2 = (220 / 187)^2$
$5000 / P_2 = 1.38$
$P_2 = 5000 / 1.38$
$P_2 = 3.6 \times 10^3$ W

Wire Gauges

Example   Find the resistance of 100 ft of No. 16 gauge
               A. aluminum wire.
               B. copper wire.

Solution

given : $L = 100$ ft ;
         No. 16 gauge wire has an area 2583 CM

A. $R = \rho L / A$
       $R = (17\ \Omega\text{-CM} / \text{ft})(100) / (2583)$
       $R = 0.66\ \Omega$

B. $R = \rho L / A$
       $R = (10.3\ \Omega\text{-CM} / \text{ft})(100) / 2583$
       $R = 0.40\ \Omega$

Solutions from Selected Problems from the Text

4. given : $i = 1.0$ A ; $t = 3.0$ s ; $q = 1.6 \times 10^{-19}$ C

    $q = i\,t$
    $q = (1.0)(3.0 = 3.0$ C
    1 electron / 1.6×10^{-19} C $= x$ electrons / 3.0 C
    $x = 1.9 \times 10^{19}$ electrons

9. given : $V_1 = 12$ V ; $i_1 = 0.25$ A ; $V_2 = 90$ V ; $I_2 = 1.88$ A

    $V_1 = SI$
    $12 = (0.25)\,(R)$
    $R = 48\ \Omega$

    $V_2 = i_2 R_2$
    $90 = (1.88) R_2$
    $R_2 = 48\ \Omega$

    Since the resistance is constant, the resistor is ohmic.

14. given : $d = (0.5\text{ in})$

$$A_{cu} = d^2$$
$$A_{cu} = (0.5 \times 10^3)^2$$
$$A_{cu} = 2.5 \times 10^5 \text{ CM}$$

19. given : $d = 0.46\text{ in}$; $L = 500\text{ ft}$; $\rho = 10.3\ \Omega\text{-CM}/\text{ft}$;

$$R = \rho L / A$$
$$R = (10.3)(500) / (4.60 \times 10^2)^2$$
$$R = 2.0 \times 10^{-2}\ \Omega$$

24. given : $R_0 = 240\ \Omega$; $T = 100°C$

$$R = R_0\ (1 + \alpha\ \Delta T) \quad R = (240)\ [\ 1 + (-0.50 \times 10^{-3})(80)\ ]$$
$$R = 230\ \Omega$$

30. given : $P = 1400\text{ W}$; $V = 120\text{ V}$; $t = 5.0\text{ min}$

    A. $P = E / t$
$$1400 = E / 300$$
$$E = 4.2 \times 10^5\text{ J}$$

    B. $P = V^2 / R$
$$1400 = 120^2 / R$$
$$R = 120^2 / 1400$$
$$R = 10.3\ \Omega$$

35. given : $P = 1200\text{ W}$; $V = 120\text{ V}$

    A. $P = iV$
$$1200 = (i)\ 120$$
$$i = 10\text{ A}$$

    B. $P = i^2 R$
$$P = (10)^2 (0.020)$$
$$P = 2.0\text{ W}$$

$$P_c / P = 2.0 / 1200 = 0.0017 \text{ or } 0.17\%$$

40. given : AWG 14 ; $R = 15\,\Omega$ $R = \rho L / A$

$L = R / (R/L)$
$L = 1.5 / 8.282$
$L = 0.181 \text{ km} = 181 \text{ m}$

Review Questions

1. What is necessary for an electric current to flow ?

A voltage or potential difference - similar to a temperature difference for heat flow.

2. Describe the electron current "flow" in a conductor and what "flows" in an alternating circuit ?

The chaotic movement of the electrons allow for only a small drift velocity for electrons (less than 1 cm/s). In an alternating current circuit, there is no net electron flow, but as a dc circuit, the electrical conduction in all parts of the circuit is almost instantaneous because the electric field is transmitted.

3. What is the unit of resistance in terms of Ohm's law ?

Since $R = V / I$, the resistance has the units of volt / ampere.

4. How does the resistance of a conductor vary with length, cross-sectional area, and temperature ?

The greater the length, the greater the resistance. The greater the cross-sectional area, the smaller the resistance. For most conductors, particularly metals, the resistance increases with increasing temperature.

5. How is the resistance of a particular model expressed ?

By the resistivity, which is a material property, and geometry or dimensions
$R = \rho L / A$.

6. What is essential neglected by expressing the area of a wire in circular mils ?

Since $A_{cm} = d^2$ and $A = \pi d^2 / 4$, where d is expressed in mils, we see that the constant $\pi / 4$ is neglected. It is really not neglected, since it is figured in the resistivity when expressed in Ω - CM / ft.

7. What happens to the temperature coefficient of a superconductor when it goes superconducting ?

It goes to zero, since $R - R_o = \alpha R_o \Delta T = 0$

8. Explain why a kilowatt-hour is a unit of energy.

Since $P = W / t$, then W Pt or work = power times time. The unit of power is the watt (W) or kilowatt (kW), and the unit of time is the second (s) or hour (h). Hence, the work or energy has units of W-s in standard units or kW-h in larger units.

9. If the voltage is doubled in a circuit (example replacing a 6 V battery with a 12 V battery), how is the available power to the circuit affected ?

It is increased by a factor of four since $P = V^2 / R$, assuming that the circuit resistance remains constant.   However, with a greater voltage there would be a greater current, $I = V / R$, and there would be more joule heating which could increase the resistance.

10. If a circuit called for more current such that you had to rewire it, how would the wire gauge be affected ?

A smaller wire gauge should be used since the smaller the wire gauge, the greater the cross-sectional area. This assumes that the same kind of wire was used.

Sample Quiz

(Remove the quiz from the book and test your knowledge of the chapter materials as though you were taking an in-class quiz. Check your answers with the key at the back of the Study Guide.)

Completion

1. If the length of a conductor is doubled, its resistance is

   _____.

2. To promote joule heating in applications such as heaters or toasters, the circuit element has a _____ resistance.

3. An increase of three in the wire gauge number_____ the cross-sectional area of the wire by _____ and _____ the resistance.

Multiple Choice

___4. The resistance of a wire depends on
   A. length
   B. temperature
   C. cross-sectional area
   D. all of the choices

___5. The I^2R loss of a circuit is given equivalently by
   A. I/V B. V^2/R C. VR D. VR^2

___6. The current in a resistor is doubled. By what factor would the power of the resistor change ?
   A. 1/4   B. 1/2   C. 2   D. 4

Problems

7. A copper wire with a diameter of 0.0050 in and 10 ft long is
   connected to a 6.0 V battery.
       A. Find the resistance of the wire.
       B. How much current would initially flow in the circuit ?
       C. How much charge would flow in the wire in 5.0 min ?
       D. Find the total energy expended in the wire in 5.0 min ?

8. An appliance has a power requirement of 800 W.   If it operates on a
   120 V source,
       A. how much current does it draw ?
       B. what is the resistance of the appliance ?

Chapter 22   Basic DC Circuits

Sample Problems

Resistances in Series and in Parallel

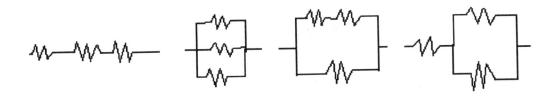

Example 1   Shown in the diagrams above are various combinations
of 5 Ω resistors. In each diagram, find the total resistance.

Solution

A. The resistors are connected in series.
$R = R_1 + R_2 + R_3$
$R = 3 \, (5 \, \Omega)$
$R = 15 \, \Omega$

B. $1/R = 1/R_1 + 1/R_2 + 1/R_3$
$1/R = 3 \, (1/5)$
$R = 1.7 \, \Omega$

C. R_s (combination in series) $= 5 \, \Omega + 5 \, \Omega = 10 \, \Omega$
$1/R = 1/5 + 1/10$
$R = 3.3 \, \Omega$

D. $1/R_p = 1/5 + 1/5$
$R_p = 2.5 \, \Omega$
$R = 2.5 \, \Omega + 5 \, \Omega$
$R = 7.5 \, \Omega$

Example 2   A 5 Ω, 10 Ω and 20 Ω resistor are connected in series to a
35 V source.
A. Calculate the total resistance.
B. Find the current in the circuit.
C. Find the voltage for each resistor.

Solution

A. $R = R_1 + R_2 + R_3$
$R = 5\ \Omega + 10\ \Omega + 20\ \Omega = 35\ \Omega$

B. $V_t = i_t R_t$
$35 = i_t (35)$
$i_t = 1.0\ A$

C. 1.0 A is the current in each resistor.

$V_5 = (1.0)(5.0) = 5.0\ V$
$V_{10} = (1.0)(10) = 10\ V$
$V_{20} = (1.0)(20) = 20\ V$

Example 3   A 3 Ω, 4 Ω, and 6 Ω resistor are connected in parallel to a 12 V source.
A. Calculate the total resistance.
B. Calculate the current and potential difference for each resistor.

Solution

A. $1 / R = 1/3 + 1/4 + 1/6$
$R = 1.3\ \Omega$

B. In a parallel combination the voltage is constant - 12 V

$V = IR$
$12 = i\ (3)$
$i_3 = 4.0\ A$

$12 = i\ (4)$
$i_4 = 3.0\ A$

$12 = i\ (6)$
$i_6 = 2.0\ A$

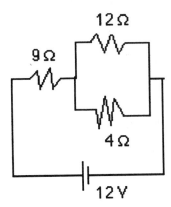

12 Ω

9 Ω

4 Ω

12 V

Example 4   The three resistors are connected as shown above to a
12 V source.
A. Find the total resistance.
B. Calculate the current for each resistor.

Solution
A.   first find the total resistance for the parallel combination :

$1/R = 1/4 + 1/12$
$R_p = 3\ \Omega$
$R = 3\ \Omega + 9\ \Omega$
$R = 12\ \Omega$

B. $V_t = i_t R_t$
$12 = i_t (12)$
$i_t = 1.0\ A$
All of the current flows in the 9 Ω resistor and through the
3 Ω equivalent resistor . Now find the voltage for the 3 Ω
equivalent.
$V_p = (1.0)(3) = 3.0\ V$

The 4 Ω and the 12 Ω resistor have the same voltage
since voltage is constant in a parallel combination.

$3 = i_4 (4)$
$I_4 = 0.75\ A$

$3 = i_{12}(12)$
$i_{12} = 0.25\ A$

Kirchoff's Rules and Multiloop Circuits

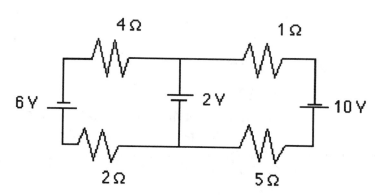

Example 1

A. Calculate the current in the 6 Ω resistor.
B. Calculate the potential difference for the 2 Ω resistor.

Solution

A. use the voltage rule :
$$0 = 10 - i\,(2) - 2 - i\,(6)$$
$$0 = 8 - 8\,i$$
$$i = 1.0\ A$$

B. $V = iR$
$$V = (2.0)(1.0)$$
$$V = 2.0\ V$$

Example 2

A. Calculate the current in the 2 Ω resistor.
B. Calculate the current in the 2 V battery.
C. Calculate the current in the 1 Ω resistor.

Solution

A. choose i_3 as moving up in the middle branch, i_1 moving counter-clockwise in the left branch, and i_2 moving clockwise in the right branch.
(solutions may vary)

$$i_3 = i_1 + i_2$$

$$0 = -4\,i_1 + 6 - 2\,i_1 + 2 \qquad 0 = -1\,i_2 + 10 - 5\,i_2 + 2$$

$$6i_1 = 8 \qquad\qquad\qquad 6\,i_2 = 12$$

$$i_1 = 1.3\,A \qquad\qquad\qquad i_2 = 2.0\,A$$

since $i_3 = i_1 + i_2$

$$i_3 = 1.3\,A + 2.0\,A$$

$$I_3 = 3.3\,A$$

Circuit Applications

Example In a slide - wire Wheatstone bridge circuit a resistor of 15 Ω is used on one side and a second 25 Ω resistor is used on the other side. Where along the meter-stick will the bridge circuit be balanced ?

Solution

given : $R_3 = 15\ \Omega$; $R_x = 25\ \Omega$

$$R_x = R_3\,(\,L_1\,/\,L_2\,)$$

$$25 = 15\,(\,L_1\,/\,L_2\,)$$

$$1.67 = L_1\,/\,L_2$$

$$1.67\,(\,1 - L_1\,) = L_1$$

$$1.67 - 1.67\,L_1 = L_1$$

$$1.67 = 2.67\,L_1$$

$$L_1 = 62\ cm$$

Solutions from Selected Problems from the Text

5. given : $R = 5.0\ \Omega$; $V = 12\,V$; $i = 3.0\,A$

$$V_p = i_p R_p$$

$$12 = (3.0)\,R_p$$

$$R_p = 4.0\ \Omega$$

$$1\,/\,R_p = 1\,/\,R_1 + 1\,/\,R_2$$

$$1\,/\,4.0 = 1\,/\,5 + 1\,/\,R$$

$$1\,/\,4 - 1\,/\,5 = 1\,/\,R$$

$$R = 20\ \Omega$$

10. first find the equivalent resistance of R_2 and R_3

$$R_{23} = R_2 + R_3$$
$$R_{23} = 10\ \Omega + 10\ \Omega$$
$$R_{23} = 20\ \Omega$$

secondly find the equivalent resistance of R_5, R_4 and R_{23}

$$1 / R_{2345} = 1 / R_5 + 1 / R_{23} + 1 / R_4$$
$$1 / R = 1 / 10 + 1 / 20 + 1 / 10$$
$$R_{2345} = 4.0\ \Omega$$

next find the equivalent resistance for R_6 and R_7

$$R_{67} = R_6 + R_7$$
$$R_{67} = 10\ \Omega + 10\ \Omega$$
$$R_{67} = 20\ \Omega$$

next find the equivalent for R_{2345} and R_{67}

$$1 / R = 1 / 4.0 + 1 / 20$$
$$R_{eq} = 20 / 6$$

Now find the total resistance

$$R_{total} = R_1 + R_8 + R_{eq}$$
$$R_{total} = 10\ \Omega + 10\ \Omega + (20 / 6)\ \Omega$$
$$R_{total} = 23.3\ \Omega$$

15.
$$1 / R_p = 1 / R_4 + 1 / R_5 + 1 / R_6$$
$$1 / R_p = 1 / 10\ \Omega + 1 / 10\ \Omega + 1 / 20\ \Omega$$
$$R_p = 4.0\ \Omega$$

A. $$i_1 = i_2 + i_3$$
$$V_s - V_3 - SI - i_2 R_2 = 0$$
$$*\ 18 - 6 - 3 i_1 - 5 i_2 = 0$$
$$V_3 - i_3 R_p + i_2 R_2 = 0$$
$$*\ 6 - 4 i_3 + 5 i_2 = 0$$
$$12 - 3 i_1 - 5 i_2 = 0$$
$$12 - 3 i_2 - 3 i_3 - 5 i_2 = 0$$

$$12 - 8 i_2 - 3 i_3 = 0$$
$$6 + 5 i_2 - 4 i_3 = 0$$

$$48 - 32 i_2 - 12 i_3 = 0$$
$$-18 - 15 i_2 + 12 i_3 = 0$$
$$30 = -47 i_2$$
$$i_2 = 0.63 \text{ A}$$
$$i_1 = 2.9 \text{ A}$$
$$i_3 = 2.3 \text{ A}$$

B. $i_3 = i_p$
$$V_p = i_p R_p$$
$$V_p = (2.3)(4) = 9.2 \text{ V}$$

$$i_4 = V_p / R_4 = 9.2 / 10 = 0.92 \text{ A}$$
$$i_6 = V_p / R_6 = 9.2 / 20 = 0.46 \text{ A}$$

20. $V_{AB} = (1/3)V = (1/3)(12) = 4.0 \text{ V}$
$$V_{BC} = ((2/3)V = (2/3)(12) = 8.0 \text{ V}$$

Review Questions

1. When resistors are connected in series, what is the voltage effect ?

    The sum of the voltage of the resistors is equal to the voltage of the battery $(V_1 + V_2 + V_3 + ... = V)$.

2. What is common to resistors connected in series ?

    The current through each resistor is the same.

3. What are the current and voltage conditions for resistors connected in parallel ?

    The current divides at a parallel junction such that the sum of the current through the resistors is equal to the input current $(I_1 + I_1 + I_3 + ... = I)$. The voltage across each resistor is the same.

4. Given several resistances, how could you get the greatest and least resistance by using all of the resistances in a circuit ?

Connecting in series would give the greatest resistance since the resistances add. Connecting in parallel would give the smallest resistance since the total resistance is less than the smallest resistance in the parallel connection.

5. Which of Kirchoff's rules reflect the conservation of charge and the law of conservation of energy ?

The current rule and the voltage rule, respectively.

6. In transversing a loop, a resistance is passed through opposite to the selected current and a battery is passed through going from the negative to the positive terminal. What signs should be used ?

By the sign convention in the text, +iR for the resistance and +V for the battery.

7. Does it make any difference in which direction (clockwise or counterclockwise) a loop is applied in Kirchoff's voltage rule ? Explain.

No, since by sign convention the same equation is obtained in both cases, with interchanges positive and negative quantities.

8. What is the effect if two batteries in series in a circuit have opposite polarity directions ?

The voltage is equivalent to the voltage difference of the batteries. In effect, the batteries work against each other and the small battery is recharged.

9. Is there current flow in a Wheatstone bridge when in a balanced or null condition ?

Yes, through the arms but not through the center connecting branch.

10. What would be the case if the rheostat shown in Fig. 22.14 were connected to the two terminals on the bands at the ends of the wire core ?

The resistance would be fixed at the maximum resistance of the rheostat.

Sample Quiz

(Remove the quiz from the book and test your knowledge of the chapter materials as though you were taking an in-class quiz. Check your answers with the key at the back of the Study Guide.)

Completion

1. For resistances in parallel, the _____ is the same for each resistor.

2. Multiloop circuits are analyzed by means of_____.

3. When three connections of a three-contact variable resistor are used, the resistor acts as a _____.

Multiple Choice

___ 4. Three resistors, 5 Ω, 10 Ω, and 15 Ω are connected in parallel with a 6.0 V battery. The voltage drop across the 10 Ω resistor is

   A. 2 V   B. 3 V   C. 6 V   D. 0

___ 5. The emf's of batteries can be measured using a
   A. Wheatstone bridge   C. rheostat
   B. voltage divider   D. potentiometer circuit

___ 6. When resistors are added in parallel, the total current will
   A. increase   C. remain the same
   B. decrease

Problems

6. Two resistors, 4 Ω and 6 Ω, are connected in series and two more
   resistors of the same value are connected in parallel. The
   arrangement are connected in series with a 12 V battery. How
   much current is drawn from the battery ?

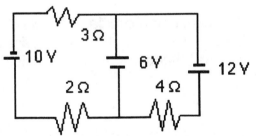

7.
   In the circuit shown above, find the current in the 2 Ω resistor, the
   3 Ω resistor, and in the 4 Ω resistor.

Chapter 23   Magnetism

Sample Problems

Electromagnetism

Example 1   A long, straight wire carries a current of 50 A. Find the magnitude of the magnetic field 50 cm from the wire.

Solution

given : $\mu_o = 4\pi \times 10^{-7}$ T-m/A ; i = 50 A ; d = 0.50 m

$B = \mu_o\, I / 2\pi d$

$B = (4\pi \times 10^{-7})(50) / 2\pi(0.50)$

$B = 2.0 \times 10^{-5}$ T

Example 2   A is bent into a circular loop with 50 turns and radius 20 cm. A current of 5.0 A is passed through the wire. Find the magnitude of the magnetic field at the center of the loop.

Solution

given : N = 50 ; I = 5.0 A ; r = 0.20 m

$B = \mu_o NI / 2r$

$B = (4\pi \times 10^{-7})(50)(5.0) / 2\,(0.20)$

$B = 7.9 \times 10^{-4}$ T

Example 3   A solenoid has 100 turns per 10 cm. The magnitude of the magnetic field produced by the solenoid is 30 mT. Find the current in the solenoid.

Solution

given : N = 100 ; L = 0.10 m ; B = 0.03 T ;

$\mu_o = 4 \pi \times 10^{-7}$ T-m/A

$B = \mu_o NI / L$

$0.03 = (4 \pi \times 10^{-7}) (100) I / 0.10$

$0.03 = (1.26 \times 10^{-3}) I$

$I = 24$ A

Example 4   Two long wires carry equal currents of 5.0 A. The wires
are separated by a distance of 30 cm.
A. Find the magnitude of the magnetic field midway
between the wires if the current in the wires in the
same direction.
B. Find the magnitude of the magnetic field midway
between the wires if the current in the wires in are in
opposite directions.

Solution

given : I = 5.0 A ; d = 0.15 m ; $\mu_o = 4\pi \times 10^{-7}$ T-m/A

$B_1 = \mu_o i / 2\pi d$

$B_1 = (4\pi \times 10^{-7})(5.0) / 2\pi(0.15)$

$B_1 = 6.7 \times 10^{-6}$ T

A. $B = B_1 + B_2$

$B = 6.7 \times 10^{-6}$ T $+ 6.7 \times 10^{-6}$ T

$B = 1.3 \times 10^{-5}$ T

B. $B = B_1 - B_2 = 0$

Magnetic Force and Torque

Example 1   In example 4 in the last section, find the force per unit
length on the wires in each part.

239

Solution

given : $\mu_0 = 4\pi \times 10^{-7}$ T-m/A ; d = 0.30 m ; l = 5.0 A

$B = \mu_0 l / 2\pi d$

$B = (4\pi \times 10^{-7})(5.0) / 2\pi(0.30)$

$B = 3.3 \times 10^{-6}$ T

$F/L = IB$

$F/L = (5.0)(3.3 \times 10^{-5})$

$F/L = 1.7 \times 10^{-5}$ N/m

Example 2 An electron has a speed of 1.0×10^3 m/s in a direction which is perpendicular to a 3.0 mT magnetic field.
A. Find the force on the wire in the magnetic field.
B. Find the acceleration of the particle in the magnetic field.
C. Calculate the radius of the electron in the magnetic field.
D. Calculate the period of the electron.

Solution

given : $v = 1.0 \times 10^3$ m/s ; $B = 3.0 \times 10^{-3}$ T ;
m = 9.1×10^{-31} kg ; q = 1.6×10^{-19} C

A. $F = qvB$

$F = (1.6 \times 10^{-19})(1.0 \times 10^3)(3.0 \times 10^{-3})$

$F = 4.8 \times 10^{-19}$ N

B. $F = ma$

$4.8 \times 10^{-19} = (9.1 \times 10^{-31})(a)$

$a = 5.3 \times 10^{11}$ m/s^2

The Ammeter and Voltmeter

Example 1 A galvonometer has a coil resistance of 10 Ω, and the coil current for full-scale deflection is 1.0 mA. The galvonometer is to be used as a 3.0 V voltmeter. Find the resistance needed.

Solution

given : $R_c = 10\ \Omega$; $V_{max} = 3.0\ V$; $i_c = 0.001\ A$;

$R_m = (V_{max} - I_c R_c) / I_c$
$R_m = [3.0 - (0.001)(10)\ ] / 0.001$
$R_m = 2990\ \Omega$

Example 2 An ammeter has a shunt resistance of 0.05 Ω for 100 mA. If the coil resistance of the galvonometer is 25 Ω, what is the coil current for full scale deflection ?

Solution

$R_s = 0.05\ \Omega$; $I_{max} = 0.1\ A$; $R_c = 25\ \Omega$

$R_s = I_c R_c / (I_{max} - I_c)$
$0.05 = I_c (25) / (0.1) - I_c$
$0.005 - (0.05)I_c = 25\ I_c$
$0.005 = 25.05\ I_c$
$I_c = 2.0 \times 10^{-4}\ A$

Solutions from Selected Problems from the Text

5. given : $\mu_o = 4\pi \times 10^{-7}\ T\text{-}m / A$; $d = 0.30\ m$; $i = 5.0\ A$

A: $B_1 = \mu_o i / 2\pi d$

$B_1 = (4\pi \times 10^{-7})\ (5.0) / 2\pi\ (0.10\ m)$

$B_1 = 1.0 \times 10^{-5}\ T$ down

$B_2 = (4\pi \times 10^{-7})\ (5.0) / 2\pi\ (0.40)$

$B_2 = 0.25 \times 10^{-5}$ up

$B = B_1 - B_2$
$B = 1.0 \times 10^{-5} - 0.75 \times 10^{-5} = 0.75 \times 10^{-5}\ T$ down

B : $B = B_1 + B_2$

        $B_1 = B_2 = (4\pi \times 10^{-7})\,(5.0)\, / \, 2\pi\,(0.15)$

        $B_1 = B_2 = 0.65 \times 10^{-5}$ T up

C : $B = B_1 + B_2$

        $B = 1.0 \times 10^{-5}$ T up

10. given : $\mu_0 = 4\pi \times 10^{-7}$ T-m / A ; $L = 0.12$ m ; $d = 3.0$ cm ; $N = 60$; $i = 5.0$ A

        $B = \mu_0 N i \, / \, L$

        $B = (4\pi \times 10^{-7})\,(60)\,(5.0)\, / \,(0.120)$

        $B = 3.1 \times 10^{-3}$ T

15. given : $i_1 = 15$ A ; $i_2 = 15$ A ; $\mu_0 = 4\pi \times 10^{-7}$ T-m / A

A. $B_1 = \mu_0 i \, / \, 2\pi d$

        $B_1 = (4\pi \times 10^{-7})(15)\, / \, 2\pi\,(0.10)$

        $B_1 = 2.0 \times 10^{-5}$ T in plane toward the reader

        $B_2 = B_1$ down

        $B^2 = B_1{}^2 + B_2{}^2$

        $B^2 = (2.0 \times 10^{-5})^2 + (2.0 \times 10^{-5})^2$

        $B = 4.2 \times 10^{-5}$ T downward at an angle of 45 ° relative to the plane

B.   NO. The fields cannot cancel out since the wires are not parallel.

20. given : $i = 2.0$ A ; $B = 5.0 \times 10^{-3}$ T ; $\varnothing = 30°$

        $F \, / \, L = IB \sin 30$

        $F \, / \, L = (2.0)(5.0 \times 10^{-3})(\sin 30)$

        $F \, / \, L = 5.0 \times 10^{-3}$ N down

25. given : coil resistance $R_c = 10\ \Omega$; $i_{max} = 3.0\ A$;

$i_c = 5.0 \times 10^{-4}\ A$

$R_s = i_c R_c / (I_{max} - I_c)$

$R_s = (5.0 \times 10^{-4})(10) / (3.0 - 5.0 \times 10^{-4})$

$R_s = 1.7 \times 10^{-3}\ \Omega$

29. given : $V_1 = 1.0\ V$; $V_2 = 3.0\ V$; $V_3 = 30$; $R_c = 5.0\ \Omega$;

$i_c = 2.5 \times 10^{-3}\ A$

1 V $R_m = (V_m - i_c R_c) / i_c$

$R_m = (1.0 - 0.0125) / (2.5 \times 10^{-3}) = 395\ \Omega$

3 V $R_m = (3.0 - 0.0125) / (2.5 \times 10^{-3}) = 1195\ \Omega$

30 V $R_m = (30 - 0.0125) / (2.5 \times 10^{-3}) = 11950\ \Omega$

Review Questions

1. What is a magnetic pole and a magnetic monopole ?

A magnetic pole is a region of apparently concentrated magnetic strength which occurs with another pole of opposite polarity. A magnetic monopole would be a single pole, which has never been found in nature.

2. What is necessary to have magnetism or a magnetic field ?

A current or movement of electrical charges.

3. If the magnetic field around a long, straight, current-carrying wire were plotted versus distance from the wire, what would be the shape of the curve ?

It would be a hyperbola since B is inversely proportional to d.

4. When a magnetic material is magnetized, what occurs in the material ?

An alignment of materials of magnetic domains.

5. What type of materials have relatively high permeabilities ? Give some examples.

Ferromagnetic materials such as iron, nickel, and cobalt.

6. If an electron moved along the right side of the page from the bottom to the top and there were a magnetic field coming out and perpendicular to the page, in what direction would the electron experience a force ?

Toward the left side of the page.

7. What would happen in a dc motor if a commutator were not used ?

The motor would not run since the coil would rotate to the position of stable
equilibrium and remain there.

8. What would happen if an ammeter were connected in parallel across a fairly large resistance ?

The current would be shortened through the meter because of its low resistance (resistances in parallel).   Since this would be almost like taking the large resistance out of the circuit, more current would flow and the ammeter might burn out.

9. What would happen if a voltmeter were connected in series instead of in parallel ?

The large resistance of the voltmeter would reduce the current flowing in the circuit.

10. How is the ampere defined physically ?

In terms of the current, which, if maintained in two long parallel wires that produces a certain force per unit length between the wires.

Sample Quiz

(Remove the quiz from the book and test your knowledge of the chapter materials as though you were taking an in-class quiz. Check your answers with the key at the back of the Study Guide.)

Completion

1. Two equivalent units of magnetic fields are _____ and _____ where the latter is the SI unit.

2. An electromagnetic is made with a material having a _____ relative permeability.

3. If a high-ohm resistance is placed in series with the coil of a galvonometer, it can be used as a _____ .

Multiple Choice

_____ 4. A moving charge in a uniform magnetic field experiences the greatest force when moving at what angle to the field ?
A. 0°   B. 45°   C. 60°   D. 90°

_____ 5. A shunt is found in a
A. dc motor   C. voltmeter
B. ammeter   D. none of the choices

_____ 6. An electron moves with a speed as shown in the magnetic field. The field is directed out of the page. Find the direction of the magnetic force.
A. into the page   C. out of the page
B. +y   D. - y

Problems

7. A long, straight, current-carrying wire has a current of 2.0 A. If an electron initially moves parallel to the wire and in the direction of the current at a distance of 10 cm from the wire,
   A. what is the magnitude of the magnetic field at the electron's initial location ?
   B. in what direction will it experience a force ?

8. If a material with a relative permeability of 100 is placed in a solenoid that without the material produces a magnetic field of 5.0 T when carrying a current of 10 A. What is the magnitude of the magnetic field with a material and the same current ?

Chapter 24 Electromagnetic Induction

Sample Problems

Faraday's Law of Induction

Example 1   A magnetic is thrust into a loop of wire, radius 20 cm. The magnetic field changes from zero to 4 mT in 2.0 s. Find the induced voltage on the coil.

Solution

given : $r = 0.20$ m ; $B_o = 0$; $B_f = 4.0 \times 10^{-3}$ T ; $t = 2.0$ s

$A = \pi r^2 = (3.14)(0.20)^2 = 0.126$ m^2

$\varepsilon = -\Delta BA / \Delta t$

$\varepsilon = -(4.0 \times 10^{-3})(0.126) / (2.0)$

$\varepsilon = 2.5 \times 10^{-4}$ V

Example 2   A 20 cm long conductor moves perpendicularly through a magnetic field of 500 μT with a constant speed of 10 m/s. Find t he induced voltage in the conductor.

Solution

given : $B = 500 \times 10^{-6}$ T ; $v = 10$ m/s ; $L = 0.20$ m

$\varepsilon = BLv$

$\varepsilon = (500 \times 10^{-6})(0.20)(10)$

$\varepsilon = 1.0 \times 10^{-3}$ V

Lenz's Law and Back Emf

Example   A dc motor has a resistance of 5.0 Ω. When connected to a voltage source it draws 8.0 A of current and after a short time draws 6.0 A of current.
A. What is the voltage source ?
B. What is the back emf when it is operating ?

Solution

given : $I_o = 8.0\ A$; $R = 5.0\ \Omega$; $I = 6.0\ A$

A. $V - \varepsilon_b = IR$
          $V - 0 = (8.0)(5.0)$
          $V = 40\ V$

B. $V - \varepsilon_b = IR$

          $40 - \varepsilon_b = (6)(5)$

          $40 - 30 = \varepsilon_b$

          $\varepsilon = 10\ V$

Electric Generators

Example 1   A 5.0 Ω resistor is connected to a 120 V outlet.
                A. Find the peak current.
                B. Find the peak voltage.
                C. Find the peak power in the resistor.

Solution

given : $R = 5.0\ \Omega$; $V_{eff} = 120\ V$;

A. $V = IR$
          $120 = I\ (5.0)$
          $I_{eff} = 24\ A$
          $I_{eff} = (0.63)I$
          $24 = (0.63)\ I$
          $I_{max} = 38\ A$

B. $V_{eff} = V_{max}\ (0.63)$
          $120 = V_{max}\ (0.63)$
          $V_{max} = 190\ V$

C. $P_{max} = I_{max}\ V_{max}$
          $P_{max} = (38)(190)$
          $P_{max} = 7220\ W$

Example 2   An alternator has 200 turns on its armature, each with an
             area of 2.0×10^{-3} m^2.   If the armature is turned in a
             20 mT magnetic field with a frequency 60 Hz,
             A. what is the maximum voltage ?
             B. what is the effective voltage ?
             C. and the alternator is connected to a load of 50 Ω, find
                the effective power and the maximum power.

Solution

            given : N = 200 ; A = 2.0×10^{-3} m^2 ; B = 0.020 T ;
                     f = 60 Hz

   A. ε_{max} = NBA (2πf)

             ε_{max} = (200)(0.020)(2.0×10^{-3}) (2π)(60)

             ε_{max} = 3.0 V

   B. V_{eff} = (0.63)V_{max}
             V_{eff} = (0.63)(3.0)
             V_{eff} = 1.9 V

   C.   V = iR
             V_{eff} = i_{eff}R
             1.9 = i_{eff} (50)

             i_{eff} = 3.8×10^{-2} A

             i_{max} = i_{eff}/ 0.63

             i_{max} = (3.8×10^{-2}) / 0.63

             i_{max} = 6.0×10^{-3} A

Transformers

Example 1 An ideal transformer has 20 turns on its primary and 100 turns on its secondary. The voltage input is 20 V and the power input is 40 W.
A. Find the current in the primary and secondary.
B. Find the voltage in the secondary.
C. Find the power in the secondary.

Solution

given : $N_p = 20$; $N_s = 100$; $V_p = 20$ V ; $P_{in} = 40$ W

A. $P = iV$
$40 = (i)\ 20$
$i_p = 2.0$ A

$N_p / N_s = i_s / i_p$
$20 / 100 = i_s / 2.0$
$i_s = 0.40$ A

B. $V_p / V_s = N_p / N_s$
$20 / V_s = 20 / 100$
$V_s = 100$ V

C. $P_{output} = i_s V_s$
$P_{output} = (0.40)\ (100)$
$P_{output} = 40$ W   (ideal transformer since $P_{in} = P_{out}$)

Example 2 A step down transformer has 100 turns on its primary and 10 turns on its secondary.
A. Find the ratio of the secondary current to primary current.
B. Find the ratio of the secondary voltage to primary voltage.

Solution

given : $N_p = 100$; $N_s = 10$

A. $i_s / i_p = N_p / N_s$
    $i_s / i_p = 100 / 10$
    $i_s / i_p = 10$

B. $V_s / V_p = N_s / N_p$
    $V_s / V_p = 10 / 100$
    $V_s / V_p = 1 / 10$

Solutions from Selected Problems from the Text

5. given : $A = 0.050 \, m^2$; $B_o = 0.75 \, T$; $B = 1.50 \, T$; $t = 1.0 \times 10^{-3} \, s$

A. $\varepsilon_o = - (\Delta B / \Delta t) \, A$

    $\varepsilon_o = - (1.5 - 0.75) (0.050) / 1.0 \times 10^{-3}$

    $\varepsilon_o = = -37.5 \, V$

B. $\varepsilon_o = -NA (\Delta B / \Delta t)$

    $\varepsilon_o = -(50)(0.05) (0.75 / 1.0 \times 10^{-3})$

    $\varepsilon_o = -1.9 \times 10^3 \, V$

10. given : $N = 20$; $A = 0.010 \, m^2$; $B = 0.75 \, Wb/m^2)$; $t = 0.30 \, s$

    $\varepsilon_o = -NA (\Delta B / \Delta t)$

    $\varepsilon_o = -(20)(0.010) (0.75 / 0.30)$

    $\varepsilon_o = -0.50 \, V$

15. given : $R_{coil} = 160 \; \Omega$; $\varepsilon_b = 90$ V ; $V = 120$ V ; $R_a = 10 \; \Omega$

A. $i_s = V / R_p = V \; (R_c + R_a) / R_c R_a$
$i_s = (120) \; (160 + 10) / (160)(10)$
$i_s = 12.8$ A

B. $i = (V - \varepsilon_b) / R$
$i = (120 - 90) / (9.4)$
$i = 3.2$ A

C. $P = \varepsilon^2 / R$
$P = 90^2 / 9.4 = 8.6 \times 10^2$ W

20. given : number of turns in primary $N_p = 100$; number of turns in secondary $= 300$;
$V_p = 120$ V ; $i_p = 6.0$ A

A. $N_s / N_p = V_s / V_p$
$300 / 100 = V_s / 120$
$V_s = 360$ V

$N_s / N_p = i_p / i_s$
$300 / 100 = 6.0 / i_s$
$i_s = 2.0$ A

B. $P_{output} = i_s V_s$
$P_{output} = (2.0)(360)$
$P_{output} = 720$ W

C. $P_{input} = i_p V_p$
$P_{input} = (6.0)(120)$
$P_{input} = 720$ W

253

25. given : $i_p = 1.0$ A ; $i_s = 6.0$ A ; $N_p = 90$

    A. $N_p / N_s = i_s / i_p$

            $90 / N_s = 6.0 / 1.0$

            $N_s = 15$

    B. $V_s / V_p = N_s / N_p$

            $V_s = V_p (15 / 90)$

            $V_s = (1/6)V_p$

Review Questions

1. Is the electromotive force (emf) really a force ? Explain.

    NO, it is the electric potential per unit charge or voltage.

2. What is necessary to induce an emf in a conducting loop ?

    Relative motion between the loop and a magnetic field such that there is a change in the number of field lines through the loop.

3. What is magnetic flux and how is it related to induced emf ?

    The magnetic field ($\phi = BA$) is a measure of the number of magnetic field lines through an area such as that of the conducting loop. If there is a time rate of change of flux through a loop, an emf is induced.

4. If Lenz's law were violated, what else would be violated ?

    The conservation of energy.

5. For a loop rotating in a horizontal magnetic field, when is the induced voltage the greatest ?

Since $\varepsilon = \varepsilon_{max} \sin \omega t$, $\varepsilon = \varepsilon_{max}$ when $\sin \omega t = 1$. This occurs when sin 90° or 90° and the perpendicular to the plane of the loop is parallel to the field.

6. If the plane of a loop is perpendicular to a uniform magnetic field and the loop is moved parallel to the field, what is the induced emf in the loop ?

Zero since there is no time rate of change of flux. The number of field lines through the loop remain constant.

7. How does the back emf in a dc motor tend to regulate the motor speed ?

The back emf depends on the rotational speed of the motor armature, and is greater with increasing speed since there is a greater change in flux. But, the back emf tends to reduce the armature current and, hence, the torque on the armature. As a result, there is some optimum operational speed.

8. What is an advantage of a universal motor ?

It operates on either ac or dc.

9. If the number of turns on a transformers greater than on the secondary, describe the operation of the transformer.

It is a step-down transformer. The voltage is stepped down by a factor of N_s / N_p and the current is stepped-up by a factor of N_p / N_s.

10. What is a major source of loss in transformers ?

Eddy currents, which are set up in the iron core as a result of the changing flux in its volume. These currents give rise to i^2R lost and lower transformer efficiency.

Sample Quiz

(Remove the quiz from the book and test your knowledge of the chapter materials as though you were taking an in-class quiz. Check your answers with the key at the back of the Study Guide.)

Completion

1. The unit of magnetic flux using the tesla is _____.

2. The direction of an induced current is given by_____.

3. The propulsion of a MagLev train is provided by a
_____.

Multiple Choice

____ 4. Assuming the outer boundary of this page were a conducting loop and the magnetic field going into the page increased, then
A. there would be no induced emf
B. there would be an induced current down the right side of the page
C. the magnetic field resulting from the induced current would be into the page
D. none of the above

____5. If the number of windings on the secondary of a transformer is greater that the number of windings on the primary, then
A. there is no mutual induction
B. the current in the primary is less than the current in the secondary
C. it is a step-down transformer
D. none of the above

Problems

6. A magnetic field changes at a constant rate of 3.5 T / s through a loop with an area of 0.25 m^2. If the loop has a resistance of 0.20 Ω, what is the initial current in the loop ?

7. A transformer has 300 windings on its primary and 1200 windings on its secondary. If 120 V is applied to the primary, and a current of 2.5 A flows, what is the voltage and current in the secondary ? (Assume this is an ideal transformer.)

Chapter 25   Basic AC Circuits

Sample Problems

Reactance

Example 1   A source of 90 V, 60 Hz, is applied to a 100 μF capacitor. Find the
A. capacitive reactance.
B. the current in t he circuit.

Solution

given : $V = 90$ V ; $f = 60$ Hz ; $C = 100 \times 10^{-6}$ F

A. $X_C = 1/2\pi fC$

$X_C = 1 / 2\pi (60) (100 \times 10^{-6})$

$X_C = 265\ \Omega$

B. $V = I X_C$
$90 = I (265)$
$I = 0.34$ A

Example 2   A source of 90 V - 60 Hz is applied to a 100 mH inductor. Find the
A. inductive reactance.
B. the current in the circuit.

Solution

given : $V = 90$ V ; $f = 60$ Hz ; $L = 100 \times 10^{-3}$ H

A. $X_L = 2\pi fL$

$X_L = 2\pi (60)(100 \times 10^{-3})$

$X_L = 37.7\ \Omega$

B. $V = I X_L$

       $90 = I \,(37.7)$

       $I = 2.4 \, A$

Impedance, Resonance, and Power Factor

Example 1   A series RC circuit has a resistance of 500 Ω and a 100 μF capacitor. If the voltage source operates at 90 V - 60 Hz, what is the

A. capacitive reactance.

B. the impedance.

C. the current flowing in the circuit.

Solution

       given : R = 500 Ω ; C = 100 x 10^{-6} F ; V = 90 V ; f = 60 Hz

A. $X_C = 1/\, 2\pi fC$

       $X_C = 1\,/\, 2\pi (60)(100 \times 10^{-6})$

       $X_C = 265 \, \Omega$

B. $Z^2 = R^2 - (X_L - X_c)^2$

       $Z^2 = 500^2 - [0 - 265 \,]^2$

       $Z = 424 \, \Omega$

C. $V = iZ$

       $90 = i \,(424)$

       $i = 0.21 \, A$

Example 2   A series RLC circuit contains a 100 Ω resistor, a 100 μF capacitor, and a 100 mH inductor. The combination is connected to a 60 V source.

A. Find the resonant frequency.

B. Find the total current.

C. Find the phase angle.

Solution

given : $R = 500\ \Omega$; $C = 100 \times 10^{-6}$ F; $L = 100 \times 10^{-3}$ H

A. from previous problems $X_L = 37.7\ \Omega$; $X_c = 265\ \Omega$

$f = 1 / 2\pi\ (LC)^{1/2}$
$f = 1 / 2\pi\ [\ (100 \times 10^{-3})(100 \times 10^{-6})]^{1/2}$
$f = 50$ Hz

B. $Z = [\ R^2 - (X_L - X_c)^2\ ]^{1/2}$

$Z = [\ 500^2 - (38 - 265)^2\ ]^{1/2}$
$Z = 446\ \Omega$
$V = I\ Z$
$60 = I\ (446)$
$I = 0.13$ A

C. $\text{Tan}\varnothing = (X_L - X_c) / R$
$\text{Tan}\varnothing = (38 - 265) / 500$
$\text{Tan}\varnothing = -0.45$
$\varnothing = -24°$

Example 3   A series RLC circuit with a 200 Ω resistor, a 20 μF capacitor, and a 0.25 H inductor is driven by a 120 V - 60 Hz source. What is the power loss of the current ?

Solution

given : $R = 200\ \Omega$; $C = 20\ \mu$F ; $L = 0.25$ H ; $V = 120$ V ;
$f = 60$ Hz

$X_c = 1 / 2\pi fC$

$X_c = 1 / 2\pi (60)(20 \times 10^{-6})$

$X_c = 133\ \Omega$

$X_L = 2\pi fL$

$X_L = 2\pi\ (60)(0.25)$

$X_L = 94\ \Omega$

$$P = IV \cos \emptyset$$
$$\cos \emptyset = R / [R^2 + (X_L - X_C)^2]^{1/2}$$
$$\cos \emptyset = 200 / [200^2 + (94 - 133)^2]^{1/2}$$
$$\cos \emptyset = 200 / 204$$
$$\emptyset = 11°$$

Electrical Power Transmission

Example   A building with a 480 V - 60 Hz service has a inductive load of 500 Ω and a resistive load of 750 Ω.
A. What is the power factor of the load ?
B. Find the capacitance needed to balance the load.

Solution
given : V = 480 V ; f = 60 Hz ; X_L = 500 Ω ; R = 450 Ω

A. $$Z^2 = R^2 + X_L^2$$
$$Z^2 = 450^2 + 500^2$$
$$Z = 673 \ \Omega$$

$$\cos \emptyset = R / \Omega$$
$$\cos \emptyset = 450 / 673$$
$$\emptyset = 48°$$

B. $$C = 1 / 2\pi f X_L$$
$$C = 1 / 2\pi (60)(500)$$
$$C = 5.3 \times 10^{-6} \ F$$

Solutions from Selected Problems from the Text

5. given : V = 100 V ; f = 50 Hz ; C = 1.5 μF

A. $$V = IX_C$$
$$X_C = 1/2\pi f C$$

$$X_c = 1 / 2\pi (50)(1.5 \times 10^{-6}) = 471 \ \mu\Omega = 2.12 \times 10^3 \ \Omega$$

$$V = IX_c$$

$$100 = I / (2.12 \times 10^3)$$

$$I = 4.7 \times 10^{-2} \ A$$

B. $I_2 / I_1 = (V_2 / V_1) \ (f_2 / f_1)$

$$I_2 / I_1 = (120 / 100)(60/50)$$

$$I_2 / I_1 = 1.4$$

10. given : $i = 0.50 \ A$; $V = 24 \ V$; $f = 50 \ Hz$

$$X_L = V / i$$

$$X_L = 24 / 0.50$$

$$X_L = 48 \ \Omega$$

$$X_L = 2\pi fL$$

$$48 = 2\pi (50)(L)$$

$$48 = 314 \ L$$

$$L = 0.15 \ H$$

15. given : $R = 150 \ \Omega$ ' $C = 7.5 \ \mu F$; $L = 500 \ mH$; $V_1 = 100 \ V$; $f_1 = 50 \ Hz$; $V_2 = 120 \ V$; $f_2 = 60 \ Hz$

$$X_C = 1 / 2\pi fC$$

$$XC_2 = 1 / 2\pi (50)(7.5 \times 10^{-6}) = 424 \ \Omega$$

$$X_{C1} = 1 / 2\pi (60)(7.5 \times 10^{-6}) = 354 \ \Omega$$

$$X_L = 2\pi fL$$

$$X_{L2} = 2\pi (50)(500 \times 10^{-3}) = 157 \ \Omega$$

$$X_{L1} = 2\pi (60)(500 \times 10^{-3}) = 188 \ \Omega$$

$$Z^2 = R^2 + (X_L - X_L)^2$$

$$Z_2^2 = 150^2 + (157 - 424)^2$$

$$Z_2 = 306 \ \Omega$$

$$Z_1^2 = 150^2 + (188 - 354)^2$$

$$Z_1 = 224 \ \Omega$$

$$i = V / Z$$
$$i_1 = 120 / 224 = 0.54 \text{ A}$$
$$I_2 = 120 / 306 = 0.39 \text{ A}$$

20. given : $R = 200 \ \Omega$; $C = 25 \ \mu F$; $V = 120 \text{ V}$; $f = 60 \text{ Hz}$

$$P = iV\cos \emptyset$$
$$\cos \emptyset = R / Z$$
$$P = iVR / Z$$
$$P = (0.53)(120)(200) / 226$$
$$\text{(this information from problem 12)}$$
$$P = 56 \text{ W}$$

25. given : $P = 100 \text{ W}$; $C = 3.0 \ \mu F$; $V = 120 \text{ V}$; $f = 60 \text{ Hz}$

$$L = 1 / 4\pi^2 f_r^2 C$$
$$L = 1 / [\ 4\pi^2 (60)^2 (3.0 \times 10^{-6}) \ ]$$
$$L = 2.3 \text{ H}$$

30. given : $V = 12000 \text{ V}$; $f = 60 \text{ Hz}$; $R = 150 \ \Omega$; $X_L = 200 \ \Omega$

    A. $Z^2 = R^2 + X_c^2$

$$Z^2 = 150^2 + 200^2$$
$$Z = 250 \ \Omega$$
$$\cos\emptyset = R / Z = 150 / 250 = 0.60$$

    B. $C = 1 / 2\pi f X_c$

$$C = 1 / 2 \ \pi \ (60)(200)$$
$$C = 1.3 \times 10^{-5} \text{ F}$$

Review Questions

1. In a pure capacitive circuit, what is the phase relationship between the current and the voltage ?

    The current leads the voltage by 90°. (ICE)

2. What is the unit of reactance ?

The ohm (Ω)

3. If the driving frequency of an RLC circuit is doubled, what are the effects on the capacitive and inductive reactances ?

The capacitive reactance is decreased by one-half and the inductive reactance is doubled. ($X_c = 1 / 2\pi\, fC$ and $X_L = 2\pi fL$)

4. When is an RLC circuit inductive ?

When the inductive reactance is greater than the capacitive reactance or the phase angle is positive.

5. When a RLC circuit is driven in resonance, what is the impedance ?

The impedance is a minimum and equal to the resistance ($Z = R$).

6. When is the power factor a maximum and when is it a minimum ?

The power factor is a maximum when the circuit is driven at resonance frequency and $\cos\phi = R / Z = R / R = 1$. To minimize the power factor, ($X_L - X_C$) must be maximized to make Z as large as possible.

7. What is the advantage of transmitting electrical power at high voltages ?

With high voltages, there are low currents and the i^2R losses in the transmission lines are less.

8. Distinguish between a ground wire and a grounding wire.

Both are at ground or zero potential, but the ground wire is part of a circuit and will carry current. A grounding wire is a dedicated ground and ordinarily does not carry current.

9. On what side of a circuit is a fuse or circuit breaker wired ?

On the high or hot side. If its were placed on the ground side and was opened, there would still be a high potential to the circuit.

10. When a three-to-two prong adapter is used, what should always be done.

The grounding wire or lug should be connected to the receptacle grounding.

Sample Quiz

(Remove the quiz from the book and test your knowledge of the chapter materials as though you were taking an in-class quiz. Check your answers with the key at the back of the Study Guide.)

Completion

1. The inductive reactance of an inductor arises because of

 _____.

2. When a RLC circuit is driven in resonance, the power factor is
 _____ and the power losses are _____.

3. Elements in household circuits, such as lamps or outlets, are
 connected in _____.

Multiple Choice

___ 4. The greater the power factor of a circuit,
    A. the greater the impedance   C. the greater the phase angle
    B. the greater the power loss   D. none of the choices

___ 5. Which of the following is an electrical safety device ?
    A. type-S fuse   C. GFI
    B. polarized plug   D. all of the choices

Problems

6. An RC circuit has a resistance of 50 Ω and a capacitance of 20 μF connected in series. The circuit is driven by a 120 V , 60 Hz source. How much current flows in the circuit ?

7. A RLC circuit has a resonance frequency of 75 Hz. If the circuit resistance and impedance are 50 Ω and 70 Ω respectively,
   A. is the circuit driven at resonance ?
   B. what percent loss occurs relative tot he resonance condition ?

Chapter 26   Electronics and Solid State Physics

Solutions from Selected Problems from the Text

3. One plate conducts each positive half cycle so that the output across either R_1 or R_2 is always positive.

9. When audio ac signals are fed into the voice coil, the currents interact with the magnetic field of the magnet, causing the voice coil to move back and forth in sympathetic vibrations, which produces the compressions and rarefactions in the air or the sound.

Review Questions

1. What is the Edison effect ?

> The flow of electron in a vacuum tube with a cathode and plate for forward bias, and no current when the plate is made negative. It forms the basis of the electron tube. This effect was observed and patented by Edison, but he did not apply it.

2. Distinguish between half-wave and full-wave rectification.

> In half-wave rectification, only every other half cycle of an ac wave is conducted such that the resulting pulses have the same polarity, but each half-wave is missing. In full-wave rectification, each half-cycle is conducted, such
> that the output has the same polarity with all half-waves present.

3. What can a vacuum tube triode do that a diode cannot ?

> Amplify

4. Distinguish between a donor and an acceptor impurity.

   A donor impurity contributes extra electrons. An acceptor impurity lacks a valance electron and creates a "hole" or vacancy in the lattice.

5. Distinguish between N-type and P-type semiconductors.

   N-type semiconductors are doped with donor impurities and have electrons as charge carriers. P-type semiconductors are doped with acceptor impurities and have positive "holes" as charge carriers.

6. What happens to the potential barrier of a junction diode when a bias voltage is applied ?

   When the diode is forward biased, the potential barrier is reduced. When the diode is reverse biased, the barrier is increased.

7. What does the light emitted from a LED result from ?

   The combining of electron and holes with the release of energy.

8. What is an IC ?

   An integrated circuit (IC), which is formed on a semiconductor "chip" or wafer.

9. Distinguish between AM and FM radio.

   AM stands for amplitude modulation and FM for frequency modulations. Sound waves are modulated on radio frequency carrier waves for transmission. The FM band has higher carrier frequencies than the AM band.

10. How is a TV picture formed ?

As a result of an electron beam of a CRT scanning photosensitive screen in response to TV signals. Color is produced by the excitation of color dots. This may be accomplished by multiple (three) or single beams.

Sample Quiz

(Remove the quiz from the book and test your knowledge of the chapter material as though you were taking an in-class quiz. Check your answers with the key at the back of the Study Guide.)

1. When the maximum current flows in a diode circuit, it is said to be

   _____.

2. A solid state triode is called a _____.

3. A low-frequency speaker is called a _____.

Multiple Choice

____5. The Edison effect depends on
   A. thermionic emission
   B. rectification
   C. reverse biasing
   D. donor impurities

____6. Which of the following has the highest frequency ?
   A. AM   B. FM   C. VHF   D. UHF

____7. A P-type semiconductor material
   A. has negative charge carriers
   B. is positive in a junction diode when forward biased
   C. is doped with a donor impurity
   D. none of the above

____8. In the detection of radio,
   A. the radio wave is mixed with the TV wave for transmission
   B. the radio wave is picked up by the antenna
   C. the signal enters the tuner
   D. the wave is demodulated

Chapter 27   Light and Illumination

Sample Problems

The Nature of Light

Example 1   The wavelength of light is 700 nm.
A. Find the frequency of the light.
B. Find the period of the light.

Solution

given : $c = 3.0 \times 10^8$ m/s ; $\lambda = 700 \times 10^{-9}$ m

A. $c = \lambda f$
$3.0 \times 10^8 = (700 \times 10^{-9}) f$
$f = 4.3 \times 10^{14}$ Hz

B. $T = 1/f$
$T = 1 / (4.3 \times 10^{14})$
$T = 2.3 \times 10^{-15}$ s

Example 2   The frequency of a radio wave is 94 MHz. Find the
wavelength of the radio wave. How long will it take the
signal to travel 100 km ?

Solution

given : $c = 3.0 \times 10^8$ m/s ; $f = 94 \times 10^6$ Hz ;
$d = 100$ km $= 1.00 \times 10^5$ m

A. $c = \lambda f$
$3.0 \times 10^8 = (94 \times 10^6) f$
$f = 3.2$ m

B. $c = x / t$
$(3.0 \times 10^8) = (1.00 \times 10^5) / T$
$T = 3.3 \times 10^{-4}$ s

272

Interference

Example 1   Two waves with t he same frequency arrive at a point .
The path lengths for the waves are 7.5 λ and 5.0 λ .
What is the phase difference for the two waves. What type
of interference is present ?

Solution

given : $d_1 = 7.5 \lambda$; $d_2 = 5.0 \lambda$

$\Delta x = 7.5 \lambda - 5.0 \lambda = 2.5 \lambda$ (path difference)

$\Delta \theta = 2\pi \Delta x / \lambda$
$\Delta \theta = 2\pi (2.5 \lambda) / \lambda$
$\Delta \theta = 5\pi$ - total destructive interference occurs.

Example 2   A monochromatic light , λ = 400 nm source shines
between two slits which are 1.0 mm apart. The image is
seen on a screen which is 2.0 m away. Find the distance
from the center of the third order bright fringe to the center
of the central maximum.

Solution

given : $\lambda = 400 \times 10^{-9}$ m ; $d = 1.0 \times 10^{-3}$ m ; $n = 3$;
L = 2.0 m

$\lambda = y_n d / nL$
$400 \times 10^{-9} = y_n (1.0 \times 10^{-3}) / (3)(2.0)$
$y_n = 2.4 \times 10^{-3}$ m

Diffraction

Example 1   Light, wavelength 500 nm shines through a diffraction
grating with 530 groves/mm. How far from the central
image will a third order bright be located on a screen
2.0 m away.

Solution

$$d = (1/530) \text{ mm} = 1.89 \times 10^{-3} \text{ m} = 1.89 \times 10^{-6} \text{ m} \; ;$$
$$\lambda = 500 \times 10^{-9} \text{ m} \; ;$$
$$y = 2.0 \text{ m} \; ; n = 3$$
$$n \lambda = d \sin \theta$$
$$(3)(500 \times 10^{-9}) = (1.89 \times 10^{-6}) \sin \theta$$
$$\theta = 52.5°$$

$$\tan \theta = x/y$$
$$\tan 52.5 = x / 2.0 \text{ m}$$
$$x = 2.6 \text{ m}$$

Example 2   A diffraction grating has 500 groves / mm. How wide is the first order image when white light passes through the diffraction grating on a screen 1.00 m away ?

Solution

$$\text{given} : \lambda_{red} = 700 \times 10^{-9} \text{ m} \; ; \lambda = 400 \times 10^{-9} \text{ m} \; ; n = 1$$
$$d = (1/500) \text{ mm} = 2.0 \times 10^{-6} \text{ m}$$

$$n \lambda = d \sin \theta$$
$$\text{red} \quad (1)(700 \times 10^{-9}) = (2.0 \times 10^{-6}) \sin \theta$$
$$\theta_{red} = 20.5°$$

$$\text{violet} \quad (2)(400 \times 10^{-9}) = (2.0 \times 10^{-6}) \sin \theta$$
$$\theta_{violet} = 11.5°$$

$$\tan \theta = x/y$$
$$x_{violet} = \tan(11.5) (1.00) = 0.20 \text{ m}$$

$$x_{red} = \tan(20.5)(1.00) = 0.37 \text{ m}$$

$$\Delta x = 0.37 \text{ m} - 0.20 \text{ m} = 0.17 \text{ m or } 17 \text{ cm}$$

Illumination

Example 1   A light has an intensity of 700 cd. Find the illumination of the light source a distance of 2.0 m from the source.

Solution

given : $I = 700$ cd ; $r = 2.0$ m

$E = I / r^2$
$E = 700 / 2.0^2$
$E = 175$ fc

Example 2   Two light sources are placed on opposite ends of a meter stick. The light placed at the 0 cm mark has an intensity of 50 cd and the light source located at the 100 cm mark has an intensity of 100 cd. At what location along the meter stick will the illumination be the same ?

Solution

given : $I_1 = 50$ cd ; $I_2 = 100$ cd ; $r_1 = x$; $r_2 = 1 - x$

$E_1 = E_2$
$I_1 / r_1^2 = I_2 / r_2^2$
$50 / x^2 = 100 / (1 - x)^2$
$7.1 / x = 10 / (1 - x)$
$7.1 - 7.1 x = 10 x$
$7.1 = 17.1 x$
$x = 0.415$ m
$x = 41.5$ cm

Solutions from Selected Problems from the Text

4. given : $v = 2.3 \times 10^8$ m/s ; $f = 6.0 \times 10^{14}$ Hz ;

A. $v = \lambda f$
$2.3 \times 10^8 = \lambda (6.0 \times 10^{14})$

$$\lambda = 3.83 \times 10^{-7} \text{ m} = 383 \text{ nm}$$

B. $\lambda_m / \lambda_v = c_m / c$

         $\lambda_m / \lambda_v = 2.3 \times 10^8 / 3.0 \times 10^8 = 0.77$

9. given : $f_1 = f_2$; $d_1 = (8.0)\lambda$; $d_2 = (7.5)\lambda$

         $\Delta\theta = 2\pi \, \Delta x = \lambda$

         $\Delta\theta = 2\pi \, (8.0 - 7.5)\lambda / \lambda$

         $\Delta\theta = 2\pi$
         this illustrates destructive interference.

14. given : $\lambda_1 = 400$ nm ; $\lambda_2 = 700$ nm ; $L = 1.2$ m ; $d = 1.0 \times 10^{-3}$ m

         $\Delta y = y_2 - y_1$

         $\Delta y = (L/d)\,(\lambda_2 - \lambda_1)$

         $\Delta y = [\, 1.2 / (1.0 \times 10^{-3})\,]\,(700 - 400)\,(1 \times 10^{-9})$

         $\Delta y = 0.36 \times 10^{-3}$ m $= 0.36$ mm

20. given : $d = (1 / 10000)$ cm $= 1.0 \times 10^{-6}$ m ; $\theta = 15°$

         $n\lambda = d\sin\theta$

         $(1)\,\lambda = (1.0 \times 10^{-6})(\sin 15)$

         $\lambda = 2.59 \times 10^{-7}$ m $= 259$ nm

26. $(e_r\text{'s})$ are from text Figure 27.17

     (550 nm) $F_e = e_r F$

                    $F_e = (1.0)(1.0) = 1.0$ lm

     (600 nm) $F_e = (0.6)(1.0) = 0.60$ lm

31. given : $E_1 = 800 \text{ lx}$; $r_1 = 2.0 \text{ m}$; $r_2 = 4.0 \text{ m}$

$$E = I/r^2 \qquad \text{(I is the same for both.)}$$

$$E_2 / E_1 = r_1^2 / r_2^2$$
$$E_2 = E_1 (r_1 / r_2)^2$$
$$E_2 = 800 (2/4)^2$$
$$E_2 = 200 \text{ lx}$$

Questions

1. What is the order of magnitude of the frequency and wavelength of visible light ?

    On the order of 10^{14} Hz and several thousand angstrom units (4000 - 7000 Å) or several hundred nanometers (40 - 700 nm).

2. What are the conditions on the phase difference and path difference for constructive and destructive interference of two waves ?

    For total constructive interference, the phase difference is zero or an even integral numbers of π's, which occur when the path difference is zero or an integral number of wavelengths. For total destructive interference, the phase difference is an odd integral number of π's, which occur when the path difference is an odd integral of half wavelengths.

3. What is the principle of Young's experiment ?

    Interference. The conditions on the path lengths for interference allows the computation of the wavelength of monochromatic light.

4. Why is $(1/4)\lambda$ film called nonreflective when on a lens, yet a $(1/4)\lambda$ soap film is reflective for the same wavelength ?

277

For a soap bubble, there is a 180° phase shift at the inner surface, so a total path length of $(1/2)\lambda$ gives constructive interference. For a film on a lens, both reflections at the film surfaces undergo 180° phase shifts, so the interference is destructive or the film is nonreflective.

5. What is diffraction ?

The bending of waves around the corner or edge of an opaque object.

6. In a polarization sheet, how is transmission axis or polarization axis related to the orientation of the molecular chains ?

The transmission axis is perpendicular to the molecular chains. The **E** vectors parallel to the chains are absorbed by the molecular electrons.

7. The majority of the light emitted from an incandescent bulb is in what region ?

Infrared. This is why a typical incandescent bulb is less than 5% for lighting.

8. What is fluorescence ?

The process whereby a material absorbs light at one wavelength and emits light at another wavelength. This is the principle of fluorescent lamp which a fluorescent material that absorbs ultraviolet light from the bulb and emits visible light.

9. Is the lumen a true physical unit ? Explain.

The lumen unit relates the luminous flux to the visual brightness perceived by the eye, and hence is not a true physical unit since it involves an average physiological response.

10. What is the light flux per area and what are its units ?

This is illuminance or illumination ($E = F/A$) and is measured in units of lumen/ft^2 or foot-candle and lumen/m^2 or lux, which is the SI unit.

Sample Quiz

(Remove the quiz from the book and test your knowledge of the chapter material as though you were taking an in-class quiz. Check your answers with the key at the back of the Study Guide.)

Completion

1. The separation of light into its component colors is called
   _____.

2. A path length difference of 2λ would give rise to
   _____ interference.

3. What is the wavelength and color of maximum visual sensitivity?
   _____ and _____

Multiple Choice

____ 4. Which of the following is not based on interference?
   A. Young's experiment   C. fluorescence
   B. diffraction   D. all of the choices

____ 5. A measure of the power or time rate of flow of light energy is
   A. luminous flux   C. illumination
   B. illuminance   D. intensity level

Problems

6. The wavelength of monochromatic light is 500 nm.
  A. What is the frequency of the light ?
  B. What is the minimum path difference between two sources of
     this light that would give total destructive interference ?

7. A diffraction grating has 9000 lines/cm. At what angle will the first
   order fringe be observed for monochromatic light with a frequency of
   7.5×10^{14} Hz ?

Chapter 28   Mirrors and Lenses

Sample Problems

Reflection and Refraction

Example 1   A ray of light makes an angle of 30° with a surface. Find the angle of reflection.

Solution

     given : the angle with the surface is 30°, therefore the
           angle of incidence
           in 60°.

     by the law of reflection the angle of incidence equals the
     angle of reflection, therefore the angle of reflection in 60°

Example 2   Find the speed of light in a piece of ice.

Solution

     from Table 28.1, the index of refraction for ice is 1.31

$$n = c / c_m$$
$$1.31 = 3.0 \times 10^8 / c_m$$
$$c_m = 2.29 \times 10^8 \text{ m/s}$$

Example 3   A ray of light is incidence at a 30° angle as it passes from air into water. Find the angle of refraction.

Solution

     given : $\theta_1 = 30°$; $n_1 = 1.00$; $n_2 = 1.33$

     $n_1 \sin \theta_1 = n_2 \sin \theta_2$

     $(1.00) \sin 30 = (1.33) \sin \theta_2$

     $0.5 / 1.33 = \sin \theta_2$

     $\theta_2 = 22°$

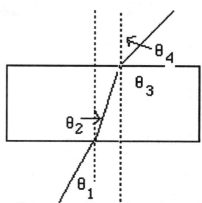

Example 4

A beam of light comes passes from water into a piece of glass and then into water as shown above. The angle of incidence $\theta_1 = 20°$. Assuming the index of refraction for glass is 1.5 and the index for water is 1.33, find θ_2, θ_3, and θ_4.

Solution

given : $n_1 = 1.33$; $n_2 = 1.5$; $n_3 = 2.0$; $\theta_1 = 20°$

$n_1 \sin \theta_1 = n_2 \sin \theta_2$

$(1.33)(\sin 20) = 1.5 (\sin \theta_2)$

$\theta_2 = 18°$

$n_2 \sin \theta_2 = n_3 \sin \theta_3$

$1.5 (\sin 18) = 1.0 (\sin \theta_3)$

$\theta_3 = 28°$

Example 5

A. Find the critical angle for a glass (n = 1.5) to air surface.
B. What happens if the angle of incidence exceeds the critical angle ?
C. Describe qualitatively what happens if the angle if incidence is less than the critical angle.

Solution

given : the critical angle is angle of incidence when the angle of refraction is 90°.

A. $n_1 \sin \theta_1 = n_2 \sin \theta_2$

      $(1.5) \sin \theta_1 = (1.0) \sin 90$

      $\theta_1 = 42°$

B.   If the angle of incidence exceeds the critical angle, the light is internally reflected.

C.   If the angle of incidence is less than the critical angle the light is refracted.

Mirrors

Example 1   A concave mirror is placed 10.0 cm in front of an object. A real image is 40.0 cm from the mirror.
      A. What is the focal length of the mirror ?
      B. What is the radius of curvature of the mirror ?
      C. Find the magnification.

Solution

    given : $d_o = 10.0$ cm ; $d_i = 40.0$ cm

A. $1/f = 1/d_o + 1/d_i$

      $1/f = 1/10 + 1/40$

      $f = 8.0$ cm

B. $r = 2f$

      $r = 2 (8.0$ cm $)$

      $r = 16.0$ cm

C. $M = - d_i / d_o$

      $M = - (40.0 / 10.0) = - 4.00$

Example 2   A 4.00 cm tall object is placed 10.0 cm in front of a convex mirror whose focal length is 20.0 cm.
      A. Find the location of the image.
      B. Find the height of the image.

Solution

given : h_o = 4.00 cm ; d_o = 10.0 cm ; f = -20.0 cm (since the mirror is convex)

A. $1/f = 1/d_o + 1/d_i$
       $1/(-20.0) = 1/10.0 + 1/d_i$
       $1/d_i = 1/(-20.0) - 1/10.0$
       d_i = -6.77 cm

B. $M = - d_i / d_o$

       M = - (-6.67) / 10.0 cm
       M = 0.667 x

       $M = -h_i / h_o$
        0.667 = h_i / 4.00 cm
       h_i = -2.67 cm

Lenses

Example 1   A 3.00 cm tall object is placed 40.0 cm in front of a
             converging lens, focal length 10.0 cm. Find the
             A. location of the image.
             B. height of the image.
             C. magnification.

Solution

given : h_o = 3.00 cm ; d_o = 40.0 cm ; f = 10.0 cm

A. $1/f = 1/d_o + 1/d_i$
       $1/10.0 = 1/40.0 + 1/d_i$
       $1/10.0 - 1/40.0 = 1/d_i$
       d_i = 13.3 cm

C. $M = - d_i / d_o$
       M = - (13.3) / (40.0 cm)
       M = 0.333

B. $M = -h_i / h_o$

      $0.333 = -h_i / 3.00$ cm

      $h_i = 1.00$ cm

Example 2   A 4.00 cm tall object is placed 10.0 cm in front of a lens. It produces a virtual image located 20.0 cm from the lens. Find the
A. focal length of the lens.
B. type lens used.
C. magnification.

Solution

      given : $h_o = 4.00$ cm ; $d_o = 10.0$ cm ; $d_i = -20.0$ cm

A. $1/f = 1/d_o + 1/d_i$

      $1/f = 1/10.0 + 1/{-20.0}$

      $f = 20.0$ cm

B.   since the focal length is positive, this means the lens is converging.

C. $M = -d_i / d_o$

      $M = -(-20.0) / 10.0$

      $M = 2.00$

Solutions from Selected Problems from the Text

4.   given : $c_m = 1.5 \times 10^{10}$ cm / s $= 1.5 \times 10^8$ m/s

      $n = c / c_m$

      $n = (3.0 \times 10^8) / (1.5 \times 10^8)$

      $n = 2.0$

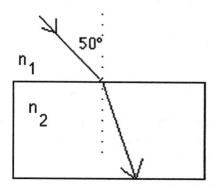

9.

given : $\theta_1 = 50°$; $n_2 = 1.50$; the water can be ignored since the material travels through the material.

A. $\sin \theta_1 / \sin \theta_2 = n_2 / n_1$

$\sin 50 / \sin \theta_2 = 1.5 / 1.0$

$\sin \theta_2 = \sin 50 / 1.5$

$\sin \theta_2 = 0.51$

$\theta_2 = 31°$

B. the same

14. given : diagram is similar to the one in the previous problem 9.
$n_2 = 1.70$; $n_1 = 1.0$

A. $n_1 \sin 45 = n_2 \sin \theta_2$

$(1.0) \sin 45 = 1.7 (\sin \theta_2)$

$\theta_2 = 24.6°$

$n_2 \sin 24.6 = n_3 \sin \theta_3$

$1.7(\sin 24.6) = 1.0 (\sin \theta_3)$

$\theta_3 = 45°$

B. $\tan \theta_2 = x / d$ (where d is the thickness)

$\tan 24.6 = x / 10 \text{ cm}$

$x = 4.6 \text{ cm}$

286

19. given : $d_O = 10$ cm ; $r = 30$ cm ;
     $r = 2f$; $f = 15$ cm

   A. $1/f = 1/d_O + 1/d_i$
       $1/15 = 1/10 + 1/d_i$
       $d_i = -30$ cm
       the (-) sign indicates the image is virtual.
       $M = -d_i / d_O$
       $M = -(-30) / 10$
       $M = + 3$ (upright)

   B. A virtual image cannot be placed on a screen.

24. given : $f = -40$ cm ; $d_O = 4.0$ cm ; $h_O = 6.0$ ft

   A. $1/f = 1/d_O + 1/d_i$
       $1/-40 = 1/4.0 + 1/d_i$
       $d_i = -3.6$ cm
       $M = -d_i / d_O$
       $M = -(-3.6 / 4.0)$
       $M = 0.90$ or 90% - this represents a 10% reduction in size.

   B. $M = y_i / y_O$
       $y_i = (0.90) (72) = 65$ in.

30. given : $d_O = 30$ cm ; $f = 20$ cm

   A. $1/f = 1/d_O + 1/d_i$
       $1/20 = 1/30 + 1/d_i$
       $d_i = 60$ cm   (real image)
       $M = -d_i / d_O$
       $M = -60/30 = -2$ (inverted)

   B. $M = y_i / y_O$
       $2 = y_i / 10$ cm
       $y_i = 20$ cm

35.   given : f = 10 cm ; d_i = 25 cm

   A.   image can be formed on a screen , therefore it is inverted.

   B. $1/f = 1/d_o + 1/d_i$
         $1/10 = 1/d_o + 1/25$
         $1/10 - 1/25 = 1/d_o$
         $d_o = 16.7$ cm

Review Questions

1. What type of reflection is occurring from the page of this book ?

   Irregular or diffuse, since it is a rough surface.

2. When light enters a different medium, what characteristic remains the same ?

   The frequency is constant. The speed and wavelength change.

3. Give **two** conditions necessary for total reflection.

   One, the angle of incidence on a medium boundary has to be greater that the critical angle for that medium, and two, the medium in which the light is traveling must have a greater index of refraction than the medium on the other side of the boundary.

4. That three optical phenomena are involved in the production of rainbows ?

   Refraction, dispersion, and internal reflection.

5. What are the image characteristics for a plane mirror ?

The image is always upright, virtual, and the same size. Also, the image
distance equals the object distance.

6. What is the focal point for a spherical mirror and how is it related to the radius of curvature of a mirror ?

The focal point is the point where rays parallel to the mirror axis focus or converge for a concave mirror and appear to emerge from a convex mirror. The focal length, or the distance to the focal point is one-half of the radius of curvature.

7. What is a virtual image ?

The simple definition is one that cannot be formed on a screen. Essentially, this means that the reflected rays diverge such that energy cannot be concentrated on a surface or screen. It appears as though the rays come from on image inside a mirror or on the object side of a lens, or a virtual image.

8. Is a virtual image formed by a concave lens ? Explain.

A virtual image is always formed by a concave or converging lens when the object is inside the focal point of the lens.

9. Distinguish between spherical aberration and chromatic aberration.

Spherical aberration occurs for mirrors and lenses because rays far from the mirror or lens axis do not pass through the focal point or plane. As a result, the image is blurred or out of focus. Spherical aberration can be reduced by reducing the effective area of the mirror or lens. Chromatic aberration in lens due to dispersion such that rays of different frequencies or colors do not converge at the same point or in the focal plane. This is corrected by using two lenses (converging and diverging) or achromatic doublet.

10. The focal length for a spherical mirror is one-half the radius of curvature ($f = R/2$).
Does this relationship hold for spherical lenses ?

NO. The focal length of a lens is given by the lens maker's equation, which involves the index of refraction of the lens. Looking at the equation, for a spherical bioconvex (converging) lens, to have $f = R/2$ would require $n = 2$, which would be high for an index of refraction.

Sample Quiz

(Remove the quiz from the book and test your knowledge of the chapter material as though you were taking an in-class quiz. Check your answers with the key at the back of the Study Guide.)

Completion

1. Light passing into a less dense medium is refracted
   _____ the normal.

2. Fiber optics makes use of the principle of_____.

3. A diverging spherical has has _____
   surfaces.

Multiple Choice

___4. We often see "heat" rising from a hot road in the summer
   because of
   A. diffuse reflection   C. aberration
   B. refraction   D. none of the choices

___5. An object is placed in front of a concave spherical mirror at a
   distance of three times the mirror's focal length. The object's
   image is
   A. virtual   C. magnified
   B. upright   D. none of the choices

___6. An image produced from a mirror is virtual and larger. What type
   of mirror is
   used ?
   A. concave   B. convex   C. plane   D. either A or B

Problems

7. A particular material has an index of refraction of 1.45.   If light is
   incident on the material in air at an angle of 35°,
        A. what is the angle of refraction ?
        B. does this exceed the critical angle of the material ?

8. An object 2.0 cm tall is placed 12 cm in front of a bioconvex lens with
   a focal length of 10 cm. Would an image be formed on a screen, and
   if so where and how tall would the image be ?

Chapter 29   Vision and Optical Instruments

Sample Problems

Projectors

Example   An image is formed a distance of 5.00 m from a projector. The slide is located approximately 40.0 mm from the lens.
A. Find the focal length of the lens.
B. Find the magnification.

Solution

given : $d_i = 5.00$ m $= 5.00 \times 10^3$ mm ; $d_o = 40.0$ mm

$1/f = 1/d_o + 1/d_i$

$1/f = 1/(5.00 \times 10^3) + 1/40.0$

$f = 4.00 \times 10^1$ mm

Microscopes

Example   A compound microscope has an eyepiece with a focal length of 20 mm and a total magnification of 30 x.   Find the focal length of the objective lens.

Solution

$M_t = 30$ x ; $f_e = 20$ mm $= 2.0$ cm

$M_t = (20)(25)$ cm^2 $/ f_o f_e$

$30 = (20)(25) / (f_o)(2.0)$

$f_o = (20)(25) / (2.0)(30)$

$f_o = 8.33$ cm

Telescopes

Example   The magnification for a refracting telescope is 100 x.   The objective lens has a focal length of 200 cm. Find the focal length of the eyepiece lens.

Solution

given : $M = 100 \, x$; $f_o = 200$ cm

$M = f_o / f_e$
$100 = 200 / f_e$
$f_e = 200 / 100$
$f_e = 2.00$ cm

Solutions from Selected Problems from the Text

2.   given : $d_o = 41.0$ mm $= 41 \times 10^{-3}$ m ; $d_i = 1.64$ m

$1/f = 1/d_o + 1/d_i$
$1/f = 1/41 \times 10^{-3} + 1/1.64$
$f = 40 \times 10^{-3}$ m $= 40$ mm

7.   given : $f_o = 15$ mm $= 1.5$ cm ; $f_e = 45$ mm $= 4.5$ cm

$M_t = (20)(25) / f_o f_e$
$M_t = (20)(25) / (1.5)(4.5)$
$M_t = 74 \, x$

12.   given : $P = 1.25$ D ; $M = 50$

$P = 1/f$
$f = 1/P$
$f = 1/(1.25)$
$f = 0.80$ m $= 80$ cm
$M = f_o / f_e$
$50 = 80 / f_e$
$f_e = 80 / 50$
$f_e = 1.6$ cm

Review Questions

1. How is the crystalline lens of the eye changed so near and far objects can be seen clearly almost instantaneously ?

   The shape or curvature of the crystalline lens is controlled by the ciliary muscles. The quick changing of the curvature and hence the focal length of the lens (called accomidation) allows sharp images to be formed on the retina.

2. What can a nearsighted person see clearly ?

   Nearby objects. Distant objects are blurred or out of focus.

3. What is color ?

   The physiological sensation of the brain in response to the light excitation of cone receptors in the retina.

4. What are the additive primaries and the subtractive primaries, and how do they differ ?

   The additive primaries are red, green and blue. The subtractive primaries are cyan, magenta, and yellow. The additive primaries are used in the addition of color production - the mixing of light. The subtractive primaries are used in the subtractive method of color production, which involves a mixing of pigments.

5. Why is the sky blue ?

   As a result of preferential of Rayleigh scattering by gas molecules of the air. The greatest scattering is in the blue end of the spectrum.

6. What kind of lens is a simple microscope ?

It is a magnifying glass or a convex or converging lens. By bringing the object inside of the focal point, a large, virtual image acts as an object for the eye.

7. The magnification for a magnifying glass is given approximately by M = (25 cm) / f. Does this mean that any magnification can be obtained by making f as small as desired ?

NO. The magnifying glass is limited to about 3x or 4x due to spherical aberrations.

8. Distinguish between the objective and ocular of a compound microscope.

The objective lens has a relative short focal length and forms a magnified image of the object. The image is formed just inside the focal point of the ocular or eyepiece lens which has a greater focal length. The eyepiece acts as a magnifying glass and further magnifies the image.

9. Distinguish between refracting telescopes and reflecting telescopes.

Refracting telescopes use lenses to collect light and produce an image. Reflecting telescopes use mirrors.

10. What is an astronomical telescope and ho can it be made into a terrestrial telescope ?

An astronomical telescope is a refracting telescope that produces an inverted image, which is no problem in astronomical work. To produce an upright image, a diverging lens is used as an eyepiece in a Galiliean telescope or a third erecting lens is used in a spyglass.

Sample Quiz

(Remove the quiz from the book and test your knowledge of the chapter material as though you were taking an in-class quiz. Check your answers with the key at the back of the Study Guide.)

Completion

1. The receptors in the human eye responsible for twilight or black and white vision are called _____.

2. If blue and yellow lights are mixed, _____ results and the colors are said to be _____.

3. A terrestrial telescope is a _____ telescope.

Multiple Choice

___ 4. The pigment sometimes referred to as "minus green" is
   A. cyan   C. yellow
   B. magenta   D. green

___ 5. In an erecting lens is used to make a terrestrial telescope, the length of the telescope is
   A. increased four times the focal length of the erecting lens
   B. decreased by one-half the focal length of the erecting lens
   C. unaffected
   D. none of the above

Problems

6. A compound microscope has an objective and an eyepiece with focal lengths of 10 mm and 15 mm, respectively.   What is the magnification ?

7. A refracting telescope has an objective and eyepiece with focal lengths of 150 cm and 5.0 cm, respectively. The telescope is made into a terrestrial scope by using an erecting lens with a focal length of 10 cm
   A. What is the magnification of the telescope ?
   B. How does the erecting lens affect the length of the telescope ?

Chapter 30   Quantum Physics and the Dual Nature of Light

Sample Problems

The Ultraviolet Catastrophe and Quantization

Example   A black-body has an energy of 1.32×10^{-30} J. Find the frequency of the oscillation.

Solution

given : $E = 1.32 \times 10^{-30}$ J ; $h = 6.63 \times 10^{-34}$ J-s

$E = hf$
$(1.32 \times 10^{-30} \text{ J}) = (6.63 \times 10^{-34}) f$
$f = 2.0 \times 10^3$ Hz

Photoelectric Effect

Example 1   The threshold wavelength for a photoelectric emission is 500 nm.
           A. Find the cutoff frequency.
           B. Find the work function.

Solution

given : $\lambda = 500 \times 10^{-9}$ m ; $c = 3.0 \times 10^8$ m/s ;
        $f = 6.63 \times 10^{-34}$ J-s

A. $c = \lambda f$
       $(3.0 \times 10^8) = (500 \times 10^{-9}) f$
       $f = 6.00 \times 10^{14}$ Hz

B. $\emptyset = hf_o$
       $\emptyset = (6.63 \times 10^{-34})(6.00 \times 10^{14} \text{ Hz})$
       $\emptyset = 3.98 \times 10^{-19}$ J

Example 2   The cutoff frequency for a photoelectric emission is
             6.6×10^{14} Hz.
             A. Find the work function.
             B. Find the kinetic energy of an electron is the incident
                light has frequency of
                (1) 400 nm.
                (2) 700 nm

Solution

given : $f_o = 6.6 \times 10^{14}$ Hz ; $h = 6.63 \times 10^{-34}$ J-s

A. $\emptyset = hf_o$

        $\emptyset = (6.63 \times 10^{-34})(6.6 \times 10^{14}$ Hz)
        $\emptyset = 4.4 \times 10^{-19}$ J

B. $K = hf - \emptyset$
        $c = \lambda f$
        $(3.0 \times 10^8) / (400 \times 10^{-9}) = 7.5 \times 10^{14}$ Hz

        $K = (6.63 \times 10^{-34})(7.5 \times 10^{14}) - (4.4 \times 10^{-19}$ J)
        $K = 5.7 \times 10^{-20}$ J

        $(3.0 \times 10^8) / (700 \times 10^{-9}) = 4.3 \times 10^{14}$ Hz

        Compare the cutoff frequency with the frequency of the
        incident light. The frequency of the incident light is less,
        therefore there is no photoelectric emission. The kinetic
        energy would be zero.

The Bohr Theory of the Hydrogen Atom

Example 1   How much energy is required to excite an electron
             A. from the first excited state to the second excited state ?
             B. from the ground state to the first excited state ?

Solution

A.   given n = 2 ; n = 3

$$E_n = -13.6 \text{ eV} / n^2$$
$$E_2 = -13.6 \text{ eV} / 2^2 = -3.4 \text{ eV}$$
$$E_3 = -13.6 \text{ eV} / 3^2 = -1.51 \text{ eV}$$
$$E = -1.51 \text{eV} - (-3.4 \text{ eV}) = 1.89 \text{ eV}$$

B. $$E_1 = -13.6 \text{ eV} / 1^2 = -13.6 \text{ eV}$$
$$E = -3.4 \text{ eV} - (-13.6 \text{ eV}) = 10.2 \text{ eV}$$

Example 2   An electron in a hydrogen atom makes a transition from the n = 4 to n = 2 state.
A. Find the energy.
B. Find the frequency of the light.

Solution

given : n = 4 ; n = 2

A. $$E_4 = -13.6 \text{ eV} / 4^2 \qquad\qquad E_2 = -13.6 \text{ eV} / 2^2$$
$$E_4 = -0.85 \text{ eV} \qquad\qquad\qquad E_2 = -3.40 \text{ eV}$$

$$E = -0.85 \text{ eV} - (-3.40 \text{ eV})$$
$$E = 2.55 \text{ eV}$$

B. $$E = 2.55 \text{ eV} = 4.08 \times 10^{-19} \text{ J}$$

$$E = hf$$
$$4.08 \times 10^{-19} = (6.63 \times 10^{-34}) f$$
$$f = 6.15 \times 10^{14} \text{ Hz}$$

X-Rays

Example   Find the cutoff wavelength of an X-ray from a tube with a voltage of 200 kV.

Solution

$$\text{given}: V = 200 \, kV = 2.00 \times 10^5 \, V$$

$$\lambda_0 = (1240 / V) \, nm$$
$$\lambda_0 = (1240 / 2.00 \times 10^5) \, nm$$
$$\lambda_0 = 6.20 \times 10^{-3} \, nm$$

Solutions from Selected Problems from the Text

5. given : $\emptyset = 1.5 \, eV \; (1.6 \times 10^{-19} \, J / 1 \, eV) = 2.4 \times 10^{-19} \, J$;
     $h = 6.63 \times 10^{-34} \, J\text{-}s$

   A. $\emptyset = hf_0$
        $2.4 \times 10^{-19} = (6.63 \times 10^{-34})(f_0)$
        $f_0 = 3.6 \times 10^{14} \, Hz$

   B. $f = f_0 = 3.6 \times 10^{14} \, Hz$

10. given : $\emptyset = 2.0 \, eV = 3.2 \times 10^{-19} \, J$; $h = 6.63 \times 10^{-34} \, J\text{-}s$;
     $f = 3.5 \times 10^{14} \, Hz$

     $E = hf$
     $E = (6.63 \times 10^{-34})(3.5 \times 10^{14}) = 2.3 \times 10^{-19} \, J$

     Since E is less than the work function there is no
      photoelectric emission. If the E were greater than W, the
      extra energy will be in kinetic energy.

15. $L_1 = h / 2\pi$
        $L_1 = (6.63 \times 10^{-34}) / 2\pi = 1.06 \times 10^{-34} \, J\text{-}s$

$$L_2 = 2L_1 = 2.12 \times 10^{-34} \text{ J-s}$$
$$L_3 = 3L_1 = 3.18 \times 10^{-34} \text{ J-s}$$

20.   given : $n = 6$; $n = 4$

$$\Delta E = 13.6 \, [\, 1/n_1{}^2 - 1/n_2{}^2)$$
$$\Delta E = 13.6 \, (1/4^2 - 1/6^2)$$
$$\Delta E = 0.472 \text{ eV} \qquad = 7.55 \times 10^{-20} \text{ J}$$
$$E = hf$$
$$7.55 \times 10^{-20} = (6.63 \times 10^{-34}) \, f$$
$$f = 1.14 \times 10^{14} \text{ Hz}$$
$$c = \lambda f$$
$$\lambda = (3.0 \times 10^8) \, / \, (1.14 \times 10^{14})$$
$$\lambda = 2627 \text{ nm}$$

25.   given : $V = 75 \text{ kV}$

$$\lambda_0 = 1240 \, / \, V$$
$$\lambda_0 = 1240 \, / \, (7.5 \times 10^3)$$
$$\lambda_0 = 0.0165 \text{ nm}$$

Review Questions

1. What was the ultraviolet catastrophe and how was it resolved ?

Classical theory predicts the intensity of cavity radiation to be proportional to the square of the radiation frequency (f^2). At high (ultraviolet) frequencies, the emitted energy would be infinitely large. Plank resolved the problem by postulating quanta of energy.

2. What characteristics of the photoelectric effect were not predicted by classical theory ?

Wave theory could not explain a cut-off frequency and the immediate photocurrent. For the latter, it would classically take an finite and relatively long time for enough energy to be supplied by waves.

3. How does quantum theory explain the cut-off frequency for the photoelectric effect ?

Since the energy of a quantum is $E = hf$, below a certain (cut-off) frequency, a photon would not have enough energy to supply the work function to free an electron.

4. What was quantitized in the Bohr theory of the hydrogen atom ?

The angular momentum, which resulted in quantitized orbit radii and energy levels.

5. What is the ground state of an atom ?

The state corresponding to a principal quantum number of $n = 1$ or the lowest state. An electron is normally in the ground state and must be excited to higher states.

6. How is the wavelength of a photon emitted from a hydrogen atom determined ?

It depends on the transition energy of the electron. Since $E = hf = hc / \lambda$, the smaller the energy differences between transition levels, the greater the wavelengths.

7. What is the principle of the laser and what does a laser do ?

Stimulated emission which can be used to amplify light.

8. What is so special about laser light ?

It is coherent - the light waves have the same frequency, phase, and direction.   Ordinary light from an incandescent lamp, is incoherent or its waves are "chaotic."

9. What are holograms ?

A film image made using reference and interference laser beams which records the information from different parts of an object "in depth." A hologram can then be used to produce three-dimensional images.

10. Why is there a cut-off frequency for an X-ray spectrum and what causes the sharp peak characteristic X-rays ?

The cut-off frequency corresponds to frequency of the emitted photon ($E = hf$) that is the total kinetic energy of an electron striking the target - the maximum frequency for the maximum energy. The characteristic X-rays are produced when an inner electron in a target atom is dislodged and the vacancy is filled by another electron and energy is emitted in the transition.

Sample Quiz

(Remove the quiz from the book and test your knowledge of the chapter material as though you were taking an in-class quiz. Check your answers with the key at the back of the Study Guide.)

Completion

1. A quantum of light is called a _____.

2. The cut-off frequency for the current of a photocell is the same as the _____ frequency of the photomaterial.

3. Since one photon gives rise to two photons in stimulated emission, this is a(n) _____ of light.

Multiple Choice

___4. The photoelectric effect was explained by
A. Plank   B. Bohr   C. Townes   D. Einstein

___5. The production of continuous X-rays is based on
A. coherence   C. electron deceleration
B. stimulated emission   D. none of the choices

Problems

6. A beam of monochromatic light with a frequency of 7.0×10^{14} Hz is incident on a photoelectric material with a work function of 2.5 eV.
   A. What is the kinetic energy of the photoelectrons ?
   B. What is the threshold frequency of the material ?

7. What is the frequency of a photon that would cause a hydrogen electron to make a transition from the ground state to the n = 3 state ?

Chapter 31

The Nucleus and Nuclear Energy

Sample Problems

The Nucleus and the Nuclear Force

Example   Find the number of protons, neutrons, and electrons in the following assuming the atom has no net charge.

A. ^3H B. ^{14}C C. ^{238}U D. ^{16}O

Solution
A.   The atomic number is 1. The number of protons and electrons is 1 and the number of neutrons is 2.

B.   The atomic number is 6. The number of protons and electrons is 6 and the number of neutrons is 8.

C.   The atomic number is 92. The number of protons and electrons is 92 and the number of neutrons is 146.

D.   The atomic number is 8 and the number of protons and electrons is 8 and the number of neutrons is 8.

Radioactivity

Example 1   Complete the following nuclear reactions.

A. $^{64}Cu \Rightarrow {}^{64}Ni +$ _____

B. $^{223}Ra \Rightarrow$ _____ $+ {}^4He$

C. $^{222}Rn \Rightarrow$ _____ $+ {}_{-1}e$

D. $^{210}Pb \Rightarrow$ _____ $+ {}^{210}Pb$

Solution

A. The mass number on the right and left side should be equal, therefore the mass number for the unknown is 0. The sum of the atomic numbers should also equal. Cu is 29 and Ni is 28 therefore the unknown is -1. The unknown is an electron $_{-1}e$.

B. $223 - 4 = 221$; $88 - 2 = 86$
^{221}Rn

C. $222 - 0 = 222$; $86 - -1 = 87$
^{222}Fr

D. $210 - 210 = 0$; $82 - 82 = 0$
No mass and no atomic number therefore gamma (electromagnetic radiation). γ

Example 2 ^{89}Kr has a half-life of 3.2 minutes. Suppose at t = 0 there are 10 kg of ^{89}Kr. How much of the isotope is left after 16 min ?

Solution

given : $t_{1/2} = 3.2$ min ; T = 16 min - 5 halflives

$2^{-5} = 1 / 32$
$(1/32) (10 \text{ kg}) = 0.31 \text{ g}$

Nuclear Reactions

Example Determine whether the following reactions are endoergic or exoergic ? Also find the energy required or energy released.

A. ^{235}U + ^{1}n $\Rightarrow$ ^{236}U
235.043925u   1.008665u   236.045563u

B. $^{210}Po \Rightarrow {}^{206}Pb + {}^{4}He$
209.98286u   205.97446u   4.002603 u

Solution

A.   add the masses on the right and compare with the masses on the left.

235.043925 + 1.008665 = 236.05259 - 236.045563 = 0.007027 u

Energy released = 0.007027 x 931.5 MeV / u = 6.55 MeV
The is an exoergic reaction.

B.   209.98286 - (205.97446 + 4.002603) = 0.005797 u
Energy = 5.40 MeV released. Process is exoergic.

Solutions from Selected Problems from the Text

5.   7 protons $= 7 (1.67265 \times 10^{-27}$ kg$)$ $= 11.71 \times 10^{-27}$ kg
     7 electrons $= 7 (9.1 \times 10^{-31}$ kg$)$ $= 0.006377 \times 10^{-27}$ kg
     7 neutrons $= 7 (1.67495 \times 10^{-27})$ $= 11.73 \times 10^{-27}$ kg

the mass of the nucleus contains protons and neutrons

$(23.436 \times 10^{-27}) / (23.4424 \times 10^{-27}$ kg$) = 99.97\%$

11.   A. $^{226}Ra \Rightarrow {}^{222}Rn + {}^{4}He$

    B. $^{60}Co \Rightarrow {}^{60}Ni + {}^{0}e$

    C. $^{210}Po* \Rightarrow {}^{210}Po + \gamma$

16. given : original activity 120 μCi ; new activity = 15 μCi

    this represent 1/8 of the original material

    $3t_{1/2} = 3\,(5.3)$
    $3t_{1/2} = 15.9$ y

21. given : $t_{1/2} = 15$ h ; activity = 3.7×10^9 decays/s

    A. $(3.7 \times 10^9)\,[\,1\text{ Ci} / (3.7 \times 10^{10}\text{ decays/s})\,]$
        = 0.10 Ci = 3.7×10^9 Bq

    B. 5 days = 120 h/15 h = $8\,t_{1/2} = nt_{1/2}$
        Activity = $N / 2^2 = (3.7 \times 10^9) / 2^8$
        = 5.6×10^4 Bq

26. A. add the masses on the left side :   199.96828 + 1.00783 =
                                                        200.97611
       add the masses on the right side: 196.96654 + 4.00260 =
                                                        200.96914

       now find the mass defect   0.00697

                                                        x 931.5

                                                        6.49 MeV
       since the mass decreases, the process is exoergic

    B. add the masses on the right : 7.01600 + 1.00783 = 8.02383
       add the masses on the left   : 4.00260 + 4.00260 = 8.0052

       now find the mass defect   0.01863
                                                  x 931.5
                                                  17.4 MeV
       since the mass decreases, the process is exoergic

Review Questions

1. Distinguish between nucleons and isotopes.

   A nucleon is a particle in the nucleus, either a proton or a neutron. An isotope is a nuclide of a particular element that has a different number of neutrons than other nuclides of the element (same number of protons).

2. Describe the nuclear force.

   It is strongly attractive, acts between any nucleon pair, and is short ranged (zero for separation distances or $r = 10^{-12}$ cm).

3. Describe the emission particles of the three types of radioactive decay.

   An alpha particle is a helium nucleus, a beta particle is an electron, and a gamma "particle" is a quantum of energy.

4. Is the radioactive decay rate linear ?

   NO. In one half-life of the nuclei of a radioactive sample of a radioactive isotope decays, in another half-life another half, and so on. This is not a linear decay, since in equal times the same number of nuclei would decay. It is a logrimithic function.

5. What must be known to use radioactive dating ?

   The half life of the radioactive isotope and the amount of or activity at some time in the past.

6. How does a Geiger counter detect or count radioactivity ?

   The energetic particle ionizes a molecule of the gas in the Geiger tube and the electron is attracted by a positive potential. The electron collides with and ionizes other atoms, and an "avalanche" current is set up momentarily. This is amplified and counted by the dector circuitry. The time for a tube to recover for another count is called dead time, which is relatively large for the Geiger counter.

7. Distinguish between endoergic and exergic reactions.

Energy is released in an exoergic reaction and energy is used in an endoergic reaction. There is a decrease and an increase in mass, respectively.

8. What must be controlled in a nuclear reactor and how is it done ?

The fission chain reaction by use of neutron-absorbing control rods. Also, the temperature or energy build-up must be controlled to prevent meltdown. This is done by a coolant system.

9. What does a breeder reactor breed ?

It "breeds" fissionable nuclear material by nuclear reactions. In effect, more fissionable material is generated that used in the chain reaction.

10. Distinguish between fission and fusion, and why is controlled fusion so difficult to achieve ?

Fission is a process of "splitting" a large nucleus into two smaller nuclei with the emission of neutrons and energy. Fusion is the fusing of two light nuclei into a heavier nucleus with the release of energy. Confinement is the problem with controlled fusion.

Sample Quiz

(Remove the quiz from the book and test your knowledge of the chapter material as though you were taking an in-class quiz. Check your answers with the key at the back of the Study Guide.)

Completion

1. Radioactivity was discovered by _____.

2. After three half-lives, what percent of a radioactive isotope would remain ? _____

3. A sustained fission chain reaction requires a _____.

Multiple Choice

___4. The mass number is conserved in
    A. alpha decay   C. gamma decay
    B. beta decay   D. these and all nuclear reactions

___5. A reactor meltdown could result from
    A. lack of critical mass   C. insertion of control rods
    B. a LOCA   D. none of the choices

Problems

6. An old bone is found to have an average of 2 beta emissions per minute per gram of carbon. Approximately how old is the bone? ($t_{1/2} = 5730$ y for C-14)

7. Complete the following reactions :

A. $^{226}\text{Ra} \Rightarrow {}^{222}\text{Rn} + \underline{\hspace{2cm}}$

B. $^{235}\text{U} + {}^{1}\text{n} \Rightarrow \underline{\hspace{1.5cm}} + {}^{94}\text{Sr} + 2\,{}^{1}\text{n}$

C. $^{27}\text{Mg} + {}^{1}\text{H} \Rightarrow \underline{\hspace{1.5cm}} + {}^{1}\text{n}$

Chapter 32 Solar Energy Technology

Sample Problems

Solar Radiation

Example
A building with a flat roof has dimensions of 50 m by 30 m. Find the average solar energy received by the building in a month.

Solution

given : $A = 50 \text{ m} \times 30 \text{ m} = 1.5 \times 10^3 \text{ m}^2$; $t = 30$ days
from the text $5.0 \text{ kWh} / \text{m}^2$ in 1 day for the central US

$E = IAT$
$E = (5.0)(1.5 \times 10^3)(30) = 2.25 \times 10^5$ kW-h

Electricity from Solar Energy

Example
If 20% of the energy absorbed in the previous example is converted to electrical energy, how much electrical energy could be produced in a month. If electricity cost $0.10 per kW-h, what is the potential for savings ?

Solution

$(0.20) (2.25 \times 10^5)$ kW-h $= 4.5 \times 10^4$ kW-h
$(4.5 \times 10^4$ kW-h$) (\$0.10 / \text{kW-h}) = \$4500$

Solutions from Selected Problems from the Text

6. given : $I = 13.5 \text{ kW/m}^2$ (50%) $= 6.75 \text{ kW/m}^2$; $A = 0.30 \text{ m}^2$; $t = 0.50$ h

$E = IAT$

$$E = (6.75 \text{ kW/m}^2)(0.30 \text{ m}^2)(0.50 \text{ h})$$
$$E = 0.10 \text{ kWh} \quad (0.239 \text{ cl} / 2.78 \times 10^{-7} \text{ kWh})$$
$$E = 8.6 \times 10^4 \text{ cal}$$

12. given : $e = 12\% ; A = 300 \text{ cm}^2$

$$E = (0.12)E_o$$
$$E = (0.12)IA$$
$$E = (0.12)(5.0 \text{ kWh/m}^2/\text{day}) (0.030 \text{ m}^2)$$
$$E = 0.018 \text{ kWh}$$

Review Questions

1. Describe the greenhouse effect.

   The selective absorption of the gases of the atmosphere, chiefly
   water vapor and carbon dioxide, of terrestrial radiation that
   heats the atmosphere and regulates the Earth's energy balance.
   It is termed "greenhouse" because the gases have similar
   absorption properties to glass.

2. Can solar energy be converted directly into electricity ?

   YES, through the use of solar cells, which are junction diodes.

3. How does thermoelectric conversion differ from solar cell electric
   conversion ?

   Solar cell conversion is the direct conversion from solar radiation
   to electricity. In thermoelectric conversion, the solar energy is
   first converted to thermal energy of a working fluid.

4. What is a heliostat ?

   A movable mirror that tracks the Sun and concentrates solar
   energy, in a central retriever system.

317

5. Distinguish between passive solar systems and active solar systems.

> In a passive system, heat energy is distributed by natural means, while in active systems, mechanical and/or electrical equipment is used.

6. What are the three ways by which radiation reaches collector surfaces ?

> Direct, scattering, and reflected.

7. What is a major drawback of solar heating ?

> It is intermittent and variable.

8. How are windows and overhangs used in passive home designs ?

> Windows are faced south (in the US) for maximum radiation transmission in the winter, and overhangs are placed over the southern-exposure windows to protect them from the sunlight in the summer.

9. What are some advantages and disadvantages of using water as a working fluid in an active solar system ?

> Water is inexpensive and has a high specific heat - relative small temperature rise for a given amount of added heat per mass. However, water could freeze and burst pipes in the cold climates in the winter.

10. Explain in general how solar radiation or heating can be used for cooling.

> Basically, the solar energy is used to do work in pumping the heat from a low temperature reservoir to a high temperature reservoir.

Sample Quiz

(Remove the quiz from the book and test your knowledge of the chapter material as though you were taking an in-class quiz. Check your answers with the key at the back of the Study Guide.)

Completion

1. Only about _____ of the insolation reaches the Earth's surface.

2. Large arrays of heliostats are used in _____ systems.

3. A thermosiphoning hot water system would be used in a _____ home design.

Multiple Choice

___ 4. In solar applications, energy reaches collector surfaces by
   A. direct radiation
   B. scattered radiation
   C. reflected radiation
   D. all of the choices

___ 5. Active solar design commonly use
   A. flat plate collectors
   B. thermosiphoning
   C. heliostats
   D. solar cells

Problems

6. If a roof has a dimensions of 20 m x 15 m, how much solar energy would it take in a week in the central United States (in joules) where the average solar intensity is 5.0 kW-h / m^2 / day ?

7. Assuming an efficiency of 20% for a type of solar cell, what would be the area of the array required to supply the energy needed to keep a 100 W light bulb burning daily ? (use the same intensity as in problem 6.)

Answers to Quizzes

Chapter 1

1. Length

2. milli

3. kilogram (kg)

4. D

5. C

6. (160 cm) (1 in. / 2.54 cm) = 63 in. = 5 ft 3 in.
   (60 kg) (2.2 lb / kg) = 132 lb

7. (a) 5.0 L
   (b) $V = w / D = 11 / 62.4 = 0.18 \text{ ft}^3$

Chapter 2

1. 8.4×10^3 and 1.6×10^{-4}

2. $\theta/2 = 1.57$ rad

3. $y = x \tan \theta$

4. B

5. D

6. $\theta = 210° = \pi + \pi/6 = 7\pi / 6$
   $= r\,\theta = (0.40)(1.67)\,\pi = 1.47$ m

7. $F_x = F_1 \cos 30 - F_2 + F_3 \cos 60$
   $F_x = 10 (\cos 30) - 8.0 + 20(\cos 60)$
   $F_x = 10.7$ N

$F_y = F_1 \sin 30 - F_3 \sin 60$

$F_y = -12.3 \, N$

$F^2 = F_x{}^2 + F_y{}^2$

$F^2 = 10.7^2 + -12.3^2$

$F = 16.3 \, N$

$\text{Tan} \, \theta = F_y / F_x$

$\text{Tan} \, \theta = 12.3 / 10.7$

$\theta = 49°$ relative tot he x-axis in the fourth quadrant

Chapter 3

1. Concurrent

2. force ; moment (lever) arm

3. torque

4. C

5. B

6. $\sum F = F_1 + F_2 = -2 \, x - y$
   Hence, $F_3 = 2 \, x + y$ so $\sum F = 0$

7. $\sum \tau = 0 = r_1 w_1 - r_2 w_2$
   and $r_2 = (w_1 / w_2) r_1 = (5.0 / 3.0) \, (20) = 33.3 \, cm$
   or at the 63.3 cm position

Chapter 4

1. velocity

2. equal ; zero

3. zero ; (since displacement is zero)

4. C

5. D

6. (a) $x = v_0 t + (1/2)at^2$

   $x = 0 + (1/2)(2.0)(5.0)^2 = 25$ m

   (b) $v_f = v_0 + at$
   $v_f = 0 + (2.0)(5.0) = 10$ m/s

7. (a) $v_f^2 = 2$ gy

   $v_f^2 = 2(32)(64)$
   $v_f = 64$ ft/s

   $y = (1/2)$ gt^2
   $64 = 1/2(32)$ t^2
   $t = 2.0$ s

8. (a) $v_f^2 = v_0^2 - 2gy$

   $0^2 = 20^2 - 2(9.8)(y)$
   $y = 20$ m ; h = 80 m

   (b) $v_f^2 = v_0^2 - 2gy$

   $v_f^2 = 20^2 - 2(9.8)(-60)$
   $v_f = -40$ m/s

   (c) $v_f = v_0 - gt$
   $-40 = 20 - (9.8)$ t
   $t = 6.1$ s

Chapter 5

1. $v_x = v_0 \cos \theta$

2. vertical or y

3. east

4. C

5. C

6. B

7. $y = (1/2)gt^2$
   $64 = (1/2)(32)t^2$
   $t = 2.0\ s$
   $v_x = x / t$
   $x = (15)\ (2.0) = 30\ ft$

8. $v_x = v_o\cos 37 = 52\ (0.8) = 42\ m/s$
   $v_y = v_o \sin 37 = 52\ (0.6) = 31\ m/s$
   A. $v_{fy} = v_{oy} - gt$
      $-31 = 31 - (9.8)\ t$
      $t = 6.3\ s$
   B. $v_f^2 = v_{oy}^2 - 2gy$
      $0^2 = 31^2 - 2\ (9.8)(y)$
      $y = 49\ m$
   C. $x = v_x t$
      $x = (42)(6.3) = 265\ m$

9. A. $a = v^2 / r$
      $25 = v^2 / 4$
      $v = 10\ m/s$
   B. $v = 2\pi r / t$
      $10 = 2\ \pi\ (4.0) / t$
      $t = 2.5\ s$

Chapter 6

1. mass

2. mass

3. action force

4. A

5. D

6. (a) $ma = F_2 - F_1$

   $(0.5) a = 6.0 - 4.0$

   $a = 4 \ m/s^2$

   (b) $x = (1/2) at^2$

   $x = (1/2) (4.0) (3.0)^2$

   $x = 18 \ m$ in the direction of F_2

7. $F = mv^2 / r$

   $F_1 = (2.0) (1.5)^2 / R = 4.5 / R$

   $F_2 = (2.0) (4.5)^2 / R = 40.5 / R$

   $F_2 = 9 F_1$

8. $m_1 a = F - m_1 g - T$

   $m_2 a = T - m_2 g$

   $4 (a) = 60 - 39.2$

   $a = 5.2 \ m/s^2$ upward

   $T = m_2 a + m_2 g$

   $T = (2.0) (5.2) + (2.0) (9.8) = 30 \ N$

9. $F = 20 \ N$ to the left

10. $m_1 a = m_1 g \sin 37 - T$

    $5 (a) = 5 (9.8) \sin 37 - T$

    $m_2 a = T - m_2 g$

    $(2.0)a = T - (2.0)(9.8)$

    $a = 1.4 \ m/s^2$

Chapter 7

1. $N\text{-}m^2 / kg^2$

2. twice the inertia

3. surface area sliding speed

4. D

5. A

6. C

7. $F = ma = GmM/r^2$
$a = GM/r^2 = (6/67 \times 10^{-11})(100)/15^2 = 3.0 \times 10^{-11}$ m/s^2

8. $f_s = \mu_s mg = (0.65)(6.0)(9.8) = 35.3$ N
$F = mg \sin 37 = (6.0)(9.8)(\sin 37) = 29.4$ N down the plane
since the frictional force is greater then the component of the
weight pulling the mass down the plane, the object will remain
at rest.

9. A. $ma = f$
$\mu = f/N$; $N = mg$
$ma = \mu mg$
$a = \mu (9.8) = (0.25)(9.8) = 2.5$ m/s^2
   B. $v_f^2 = v_o^2 + 2ax$
$0^2 = 10^2 + 2(-2.5)(x)$
$x = 20$ m

Chapter 8

1. zero

2. work - energy theorem

3. less than 1 or less than 100%

4. B

5. C

6. B

7. A. $W = mgh = (2.0)(9.8)(10) = 196$ J
    B. $W = \Delta K = (1/2)\,mv^2 = 196$ J
        $(1/2)(2.0)\,v^2 = 196$
        $v = 14$ m/s

8. $P_{out} = (Eff)\,P_{in} = (0.75)(6.0) = 4.5$ hp $= 2475$ ft-lb/s

    $W = Pt = (2475)(600) = 1.5 \times 10^6$ ft-lb

9. A. $W = Fd$
        $W = \Delta K$
        $F = ma$
        $a = F/m = 10/5.0 = 2.0$ m/s^2
    C. $v_f = v_o + at$
        $v_f = 0 + (2.0)(5.0) = 10$ m/s

    B. $\Delta K = (1/2)(5.0)(10)^2 = 250$ J

Chapter 9

1. contact time

2. mv

3. increasing (momentum is not conserved)

4. B

5. C

6. $I = m\Delta v = (0.010)\,(4 + 3) = 0.070$ N-s

7. B. $v_2 = 2m_1 v_1 / (m_1 + m_2)$
        $v_2 = 2(0.25)(10)/1.0 = 5.0$ m/s
        $m_1 v_{1o} = m_1 v_1 + m_2 v_2$
        $(0.25)(10) = (0.25)\,v_1 + (0.75)(5.0)$
        $v_1 = -5.0$ m/s

A. $P_1 = m_1 v_1 = (0.25)(-5.0) = -1.25$ kg-m/s

$P_2 = m_2 v_2 = (0.75)(5.0) = 3.75$ kg-m/s

C. 2.50 kg-m/s

8. $m_1 v_{10} = (m_1 + m_2) v$

$(1.0)(5.0) = (7.0) v$

$v = 0.71$ m/s

Chapter 10

1. displacement (distance) , velocity (speed) , and acceleration

2. slips

3. conservation of angular momentum (gyroscope principle)

4. D

5. B

6. $\tau = I\alpha = I (\omega_f^2 - \omega_0^2) / 2\theta = (3.0) [(5.5)^2 - (2.5)^2] / 2 (4\pi)$

$\tau = 2.9$ m-N

7. $(1/2)Mv^2 + (1/2) I\omega^2 = Mgh$

$(1/2)Mv^2 + (1/2) (1/2) MR^2 (v/r)^2 = Mgh$

$v^2 = (4/3) gh$

$v = 5.1$ m/s

Chapter 11

1. d_i / d_o

2. AMA / IMA

3. pitch (distance)

4. D

5. A

6. AMA = F_o / F_i = 350 / 200 = 1.75
   IMA = 1 / sin30 = 2.0
   Eff = AMA / IMA = 1.75 / 2.0 = 0.875 = 87.5%

7. $\omega_o = (D_i / D_o) \omega_i$

   ω_o = (2.0 / 4.0) (150) = 75 rpm (speed reduction)

   ω_o = (5.0 / 4.0) (150) = 187.5 rpm

Chapter 12

1. slope

2. tensile or ultimates strength

3. bulk modulus

4. B

5. C

6. F = Y ($\Delta L / L_o$) / A = Y ($\Delta L / L_o$) / ($\pi d^2 / 4$)

   F = 4 (7.0×10^{10}) (5×10^{-5}) / π (1×10^{-3})^2
   F = 4.5×10^{12} N/m^2
   This force exceeds the tensile strength so the wire could not
   be stretched by 0.50 percent.

7. k = 1 / B = ($\Delta V / V_o$) / ρ

   k = (6.0×10^{-2}) / (2.0×10^{3}) = 3.0×10^{-5} in^2 / lb

Chapter 13

1. parallel

2. amplitude and phase constant

3. (linear) mass density

4. A

5. B

6. A. 2.0 cm

   B. $\omega = 5.0 \text{ s}^{-1}$
   $\omega = 2\pi f$
   $f = 5.0 / 2\pi$
   $f = 0.80 \text{ Hz}$

7. With three nodes (one at each end and one in the middle)
   $\lambda = (2/3) \text{ L}$
   $\lambda = (2/3) (2.0) = 1.3 \text{ m}$
   $v = \lambda f$
   $f = 400 / 1.3 = 308 \text{ Hz}$

Chapter 14

1. solids

2. zero

3. decrease

4. A

5. D

6. $IL = IL_2 - IL_1 = 10 [\log (10^{-7} / 10^{-12}) - \log (10^{-4} / 10^{-12})$

   $IL = 10 [\log (10^5) - \log (10^8)]$
   $IL = - 30 \text{ dB}$

7. $v = 331 + 0.6 T_C = 331 + 0.6 (38) = 354 \text{ m/s}$
   $f' = f_s [v / (v - v_s)]$
   $f' = 400 [354 - (354 - 25)]$
   $f' = 430 \text{ Hz}$

8. $\lambda = 4L = 2(0.80) = 1.6$ m
   $v = \lambda f$
   $340 = (1.6) f$
   $f = 2.0 \times 10^2$ Hz

Chapter 15

1. adhesive, surface tension

2. absolute pressure, complete vacuum

3. less

4. C

5. D

6. (a) $p_2 - p_1 = \rho\, g\, \Delta h = (1.0 \times 10^3)(9.8)(0.10) = 9.9 \times 10^2$ N/m^2

   (b) $F_b = \rho\, g\, V_f = (1.0 \times 10^{-3})(9.8)(1 \times 10^{-3}) = 9.8$ N
      (Note: The volume of water in 1.0 L, so it has a mass of
      one kilogram which weighs 9.8 N.)

7. $A_2 = 2A_1$ and $v_2 = (A_1 / A_2)\, v_1 = v_1 / 2$. Also $h_2 = h_1 / 2$
   By Bernoulli's principle,
   $\Delta p = p_2 - p_1 = \rho\, g\, (h_1 - h_2) + (1/2)\, \rho\, (v_1{}^2 - v_2{}^2)$
   $\Delta p = \rho\, g\, (h_1 - h_1 / 2) + (1/2)\, \rho\, [v_1{}^2 - (v_1 / 2)^2]$
   $\Delta p = \rho\, g\, (h_1 / 2) + (3 / 8)\, \rho\, v_1{}^2$
   The pressure difference is equal to one-half the original
   potential energy per volume plus three-fourths of the original
   kinetic energy per unit volume.

Chapter 16

1. 20°C ; 68°F

2. method of mixtures , energy

3. triple point

4. D

5. B

6. $T_1 = -10° + 273 = 263$ K
   $T_2 = 200° + 273 = 473$ K
   $U_2 = (T_2 / T_1) U_1 = (473 / 262) (U_1 = (1.8) U_1$

7. $\Delta Q_1 = mc\Delta T = (100)(0.60) (78°-20°)=3.48 \times 10^4$ cal$=3.48$ kcal
   $\Delta Q_2 = mL_v = (100)(204) = 2.04 \times 10^4$ cal $= 20.4$ kcal
   $\Delta Q = \Delta Q_1 + \Delta Q_2 = 55.2$ kcal

Chapter 17

1. radiation

2. greater

3. heat of combustion

4. C

5. D

6. R-val = L / k and L = k(R-16) = (0.25)(16) = 4.0 in

7. $\Delta T = (\Delta V / V_0) / \beta = (\Delta V / V_0) / 3 \alpha$
   $\Delta T = (10^{-3}) / 3(3.0 \times 10^{-5}) = 11°C$

Chapter 18

1. isometric

2. (absolute) temperature

3. first

4. D

5. B

6. $\Delta S = \Delta Q / T = ML_v / T$

$\Delta S = (20)(540) / 373 = 29 \text{ cal} / K$
$\Delta S = (29 \text{ cal} / K)(4.2 \text{ J} / \text{cal}) = 122 \text{ J}$
The entropy increases (more disorder).

7. (a) $E_{th} = 1 - (Q_c / Q_h) = 1 - (58 / 100) = 0.42 = 42\%$
   (b) $E_c = 1 - (T_c / T_h) = 1 - (353 / 673) = 0.48 = 48\%$

Chapter 19

1. toward the negative charge or away from the positive charge

2. electric field (N/C)

3. volts

4. C

5. B

6. (a) $E_t = 2E = 2kq / r^2 = 2(9.0 \times 10^9)(6.0 \times 10^{-5}) / (3)^2$

   $E_t = 1.2 \times 10^5$ N/C in the positive **x** direction
   (b) $V_t = 0$

7. $U = eV = (1)(100) = 100 \text{ eV} (1.6 \times 10^{-19} \text{ J} / \text{eV}) = 1.6 \times 10^{-17} \text{ J}$

Chapter 20

1. decreased

2. with an without

3. parallel

4. A

5. C

6. $C = \varepsilon_o A / d = (8.85 \times 10^{-12})(0.50) / (1.0 \times 10^{-3}) = 4.4 \times 10^{-9}$ F

   (a) $U = (1/2)CV^2 = (1/2)(4.0 \times 10^{-9})(12)^2 = 3.2 \times 10^{-7}$ J

   (b) $C = Q/V$; $Q = (4.4 \times 10^{-9})(12) = 5.3 \times 10^{-8}$ C

   (c) found in the beginning of the problem.

7. $1/C_s = 1/C_1 + 1/C_2$; $C_s = 1.33$ µF

   $\tau = RC = (5.0 \times 10^5)(1.33 \times 10^{-6}) = 0.67$ s

   $Q = Q_{max}(e^{-t}/RC)$; $0.67 = e^{-t}/0.67$; $t = 0.27$ s

Chapter 21

1. increased

2. small

3. decreases, one-half, doubles

4. D

5. B

6. D

7. A. $R = \rho L / A = (1.7 \times 10^{-8})(10) / \pi(1.0 \times 10^{-3})^2 = 5.4 \times 10^{-2}$ Ω

   B. $V = iR$; $i = 6.0 / (5.4 \times 10^{-2}) = 1.8 \times 10^2$ A

   C. $q = it = (1.8 \times 10^2)(300) = 5.4 \times 10^4$ C

D. $P = V^2 / R = 6^2 / (5.4 \times 10^{-2}) = 6.7 \times 10^2$ W

$P = E / t$; $E = Pt = (6.7 \times 10^2)(300) = 2.0 \times 10^5$ J

8. A. $P = iV$

$800 = i \, (120)$

$i = 6.7$ A

B. $V = iR$

$120 = (6.7) R$

$R = 18 \, \Omega$

Chapter 22

1. voltage (drop)

2. Kirchoff's rules

3. voltage divider

4. C

5. D

6. A

7. $R_s = R_1 + R_2 = 4 \, \Omega + 6 \, \Omega = 10 \, \Omega$

$1/R_p = 1/R_1 + 1/R_2$

$1/R_p = 1/4 + 1/6$

$R_p = 2.4 \, \Omega$

$I = V / R = 12 / 12.4 = 0.97$ A

7. Using Kirchoff's Rules ; I_1 in the left ; I_2 in the middle ; I_3 in the right

$I_1 + I_2 = I_3$

$0 = -3 \, (I_1) + 10 - 2(I_1) + 6$

$16 = 5 \, (I_1)$

$I_1 = 3.2$ A

$0 = 12 - 3 \, I_3 + 6$

$I_3 = 2$ A

$I_2 = 0.2$ A

Chapter 23

1. Wb/m^2 and T (tesla)

2. high or large

3. voltmeter

4. D

5. B

6. D

7. (a) $B = \mu_0 I / 2\pi d = (4\pi \times 10^{-7})(2.0) / (2\pi) (0.10) = 4.0 \times 10^{-6}$ T
   (b) away from the wire

8. $B_m = K_m B$
   $B_m = (100)(5.0) = 500$ T

Chapter 24

1. $T\text{-}m^2$

2. Lenz's law

3. (linear) synchronous

4. D

5. D

6. $\varepsilon = \Delta BA / \Delta t = (3.5)(0.25) = 0.875$ V ;
   $i = V / R = (0.875) / 0.20 = 4.4$ A

7. $V_p / V_s = N_p / N_s$
$V_s = (1200 / 300) (120) = 480 \text{ V}$
$I_p / I_s = N_s / N_p$
$I_s = (300 / 1200) (2.5) = 0.63 \text{ A}$

Chapter 25

1. self-inductance

2. one- maximum (resistive)

3. parallel

4. B

5. D

6. $X_c = 1/2\pi fC = 1 / [\, 2\pi \, (60)(20 \times 10^{-6})] = 133 \; \Omega$
$Z^2 = R^2 + (-X_c)^2$
$Z^2 = 50^2 + (-133)^2$
$Z = 142 \; \Omega$
$I = V / Z = 120 / 142 = 0.85 \text{ A}$

7. (a) NO, $\cos \emptyset = R / Z = 50/70 = 0.71$, and in resonance $\cos \emptyset = 1$
(b) $P / P_r = IV \cos\emptyset / IV = \cos \emptyset = 0.71 = 71\%$

Chapter 26

1. saturated

2. transistor

3. woofer

4. very high frequency

5. A

6. D

7. B

8. D

Chapter 27

1. dispersion

2. total constructive

3. 5500 Å and yellow-green

4. C

5. A

6. (a) $v = \lambda f$
     $f = (3.0 \times 10^8) / (500 \times 10^{-9}) = 6.0 \times 10^{14}$ Hz
   (b) $\lambda / 2 = 500 / 2 = 250$ nm

7. $d = 1 / N = 1.1 \times 10^{-6}$ m ;
   $\lambda = c / f = (3.0 \times 10^8) / (7.5 \times 10^{14}) = 4.0 \times 10^{-7}$ m or 750 nm

Chapter 28

1. away from

2. internal reflection

3. concave

4. B

5. D

6. (a) $\sin\theta_2 = \sin\theta_1 (n_1 / n_2)$

        $\sin\theta_2 = \sin 35 (1 / 1.45)$

        $\theta_2 = 23\,°$

  (b) $\sin\theta_2 = \sin\theta_1 (1 / 1.45)$

        $\theta_2 = 44°$; so it does not exceed

7. $1 / f = 1/d_o + 1/d_i$

  $1/10 = 1/12 + 1/d_i$

  $d_i = 60$ cm (real image)

  $y_i = M\, y_o = (60/12)\,(2.0) = 10$ cm (inverted)

Chapter 29

1. rods

2. white, complementary

3. refracting

4. B

5. A

6. $M_t = (20)(25)$ $cm^2 / f_o\, f_e = 500 / (1.0)(1.5) = 333$ x

7. (a) $M = f_o / f_e = 150 / 5.0 = 30$ x (Erecting lens has a magnification of one so makes no difference.)

  (b) $\Delta L = 4\,f = 4\,(10) = 40$ cm

Chapter 30

1. photon

2. threshold

3. amplification

4. D

5. C

6. (a) $\phi = (2.5 \text{ eV}) (1.6 \times 10^{-19} \text{ J / eV}) = 4.0 \times 10^{-19} \text{ J}$

   $K = hf - \phi = (6.63 \times 10^{-34})(7.0 \times 10^{-14}) - (4.0 \times 10^{-19})$
   $K = 6.0 \times 10^{-20} \text{ J}$
   (b) $f_o = \phi / h = (6.0 \times 10^{-19}) / (6.63 \times 10^{-34}) = 6.0 \times 10^{14} \text{ Hz}$

7. $\Delta E = 13.6 (1/n_f^2 - 1/n_i^2) = 13.6 (1/9 - 1) = -12.1 \text{ eV}$
   (minus because of energy input)
   $\lambda = 12{,}400 / E = 12{,}400 / 12.1 = 1024.8 \text{ Å} = 1.0248 \times 10^{-7} \text{ m}$
   $f = c / \lambda = (3.0 \times 10^8) / (1.0248 \times 10^{-7}) = 2.9 \times 10^{15} \text{ Hz}$

Chapter 31

1. Becquerel

2. 12.5%

3. critical mass

4. D

5. B

6. $3t_{1/2} = 3 (5730) = 17190 \text{ y}$

7. (a) ^4He (b) ^{140}Xe (c) ^{27}Al

Chapter 32

1. 50%

2. central receiver

3. passive

4. D

5. A

6. $E = IAt = (5.0 \text{ kW-h/m}^2/\text{day}) (300 \text{ m}^2) (7 \text{ days}) = 10,500 \text{ kW-h}$
   $(10,500 \text{ kW-h}) (3.6 \times 10^6 \text{ J/kW-h}) = 3.78 \times 10^{10} \text{ J}$

7. (Intensity)(Area)(efficiency) $= 100 = 0.10 \text{ kW} = E$
   or $A = E_o / I (\text{eff}) = (0.10 \text{ kW}) / (0.21 \text{ kW/m}^2) (0.20) = 2.4 \text{ m}^2$